Earths of Distant Suns

Michael Carroll

Earths of Distant Suns

How We Find Them, Communicate with Them, and Maybe Even Travel There

Copernicus Books is a brand of Springer

Michael Carroll
Fellow, International Association of Astronomical Artists
Littleton, CO, USA

ISBN 978-3-319-43963-1   ISBN 978-3-319-43964-8   (eBook)
DOI 10.1007/978-3-319-43964-8

Library of Congress Control Number: 2016951720

Printed on acid-free paper

This Copernicus imprint is published by Springer Nature
The registered company is Springer International Publishing AG
The registered company address is: Gewerbestrasse 11, 6330 Cham, Switzerland

Acknowledgments

My thanks to Babak Tafreshi, talented astrophotographer, for the lovely view of the ALMA telescope. Babak, you let us hear the heavens sing! Nick Stevens played a key part in my understanding of interstellar ships and nuclear pulse propulsion. As if that weren't enough, he shared his artistic masterpieces for the book! Steve Walker and Dorian Houser of the National Marine Mammal Foundation put up with my wacky questions with good humor. So did my buddy, "dolphin whisperer" Alan Lewis, who got them into it in the first place. Thanks, Alan! Adrian Brown and Matthew Tiscareno played instrumental roles in my visit to SETI, as Chris McKay and Elisa Quintana did for my tour in NASA/Ames' astrobiology wonderland. (Thanks for your office, Chris!) Special commendation in the "Twilight Zone" category goes to premiere science writer Tina Saey, who made the same bizarre Star Trek/silicon life connection I did. Her article for *Science News* came out the week after I submitted this manuscript to my line editor. Good thinking, Tina! The first reader and line/content editor extraordinaire Marilyn Flynn lent her magic. Finally, my gratitude to Caroline, who not only organized this literary mess but always lends structure and meaningful organization to my own wacky life. Love you, sweetie!

Contents

About the Author

Author/artist Michael Carroll has spent decades as a science journalist and even longer as an astronomical artist. He received the AAS Division of Planetary Science's Jonathan Eberhart Award for the best planetary science feature article of 2012. He lectures extensively in concert with his various books and has done invited talks at science museums, aerospace facilities, and NASA centers. He has written articles and books on topics ranging from space to archeology. His articles and art have appeared in *TIME*, *National Geographic*, *Scientific American*, *Smithsonian*, *Popular Science*, *Astronomy*, *Sky and Telescope*, *Astronomy Now (UK)*, and a host of children's books and magazines. Among his twenty-some books are Springer's *Living Among Giants: Exploring and Settling the Outer Solar System* and his novel *On the Shores of Titan's Farthest Sea* for Springer's Science and Fiction series (2015). One of his paintings is on the surface of Mars—in digital form—aboard the Phoenix lander. Carroll is the 2006 recipient of the Lucien Rudaux Award for lifetime achievement in the astronomical arts. He is a fellow and founding member of the International Association of Astronomical Artists.

1

A Place Like Home

But who shall dwell in these Worlds if they be inhabited? … Are we or they Lords of the World? … And how are all things made for man?

—Johannes Kepler

Our Solar System used to be a lonely, forlorn place. It spun along the Orion arm of the Milky Way Galaxy, trailing nine (yes, *nine*) planets, a flotilla of asteroids and comets, and one singular planet encrusted with something special: life. Beyond lay a myriad of stars, but did any of them have life forms? Our best telescopes could not resolve planets orbiting even the nearest of suns. Was it possible that Earth was alone, or at the very least a very, very rare case?

Something within human nature causes us to yearn for community, or at least engenders a captivation with *the other*. Seth Shostak is the Senior Astronomer and Director of the SETI Research Institute. "I think we're hard-wired to be interested in that possibility," he surmises. "You go into any classroom, and say, 'How many of you kids think there are aliens out there?' Essentially all the hands go up. You can say that it's because they've seen them on TV and in the movies. That's true, but the reason that movies or TV portray these guys is that the audience accepts that idea. The reason that they're interested in it is the same reason all kids are interested in dinosaurs. We're hard-wired to be interested in predators. Anything with big teeth is of interest; you want to pay attention to what their habits are or you're out of the gene pool. *Animal Planet* does stories about predators. They know their audience is not so interested in goldfish. Sharks are better than gerbils." The concept of aliens, he says, may provide a similar psychological situation, a mindset that

© Springer International Publishing Switzerland 2017
M. Carroll, *Earths of Distant Suns*,
DOI 10.1007/978-3-319-43964-8_1

ties into simple survival. "It seems logical that you'd be paying attention to the possibility of potential competitors or mates. If there's another tribe on the other side of the hill, it's important to know about those guys. They might provide an opportunity, but on the other hand, they may come over the hill and take all your land or just kill you."

Throughout history, the question, "Are we alone in the universe?" has had philosophical and theological as well as scientific ramifications (see Chap. 2 for details). Ancient Jewish writings from the book of Genesis[1] seemed to indicate that Earth was a special creation, and as such, it was unlikely that other worlds would have life. The idea was not held by all religious people. Albertus Magnus, John Calvin and many other theologians saw the cosmos as an expression of a creative God. From their perspective, a few extra planets—and perhaps a few more beings—could only add to the wonder of the Creator.

The "plurality of worlds," the idea that there are many Earths out there, was not a new one. Some ancient Greek thinkers saw the universe as arranged in an echoing and repeating pattern: an Earth at center, surrounded by a sun and planets. It was a view held by the Pythagoreans, Democritus and Leucippus, the Stoics Epicurus and Lucretius, and others. It was believed that these Earth-centered universes operated independently of each other, each arranged as concentric crystalline spheres that held various planets or stars embedded upon their inner surfaces for we Earth dwellers to see. Ancient writings did not focus on the concept of communicating with such remote places.

The idea of other living worlds had its proponents and skeptics on the societal level as well. Pluralism's influence reached into the Islamic golden age as early as the A.D. 700s. One of the stories from the *One Thousand and One Arabian Nights*, a collection of tales from the eighth to the thirteenth centuries, describes the "Adventures of Bulukiya." While on a quest to find an herb of immortality, the protagonist visits Heaven and Hell, and travels to worlds beyond Earth, worlds inhabited by a variety of exotic life forms. Another tale, the eleventh-century adventure *Salaman wa-Absal,* could have been written by a nineteenth-century science fiction author. The hero, Absal, leaves Earth to visit sentient beings on the Moon and several planets, then goes on to explore the stars, where he meets the dwellers of great "zodiacal cities."

The plurality of worlds concept was resurrected in the "machine age" of the Industrial Revolution. In 1817, Scottish lecturer Thomas Chalmers submitted to his audiences that the planets "must be the mansions of life and intelligence…there are other worlds, which roll afar; the light of other suns shines upon them; and the sky which mantles them is garnished with other

---

[1] Genesis is also accepted as a holy book by Christians and Muslims.

stars."[2] Other orators followed in his footsteps, adding to the idea of Earth as one among many. In 1827, philosopher and theologian William Whewell preached, "Earth is one among a multitude of worlds…with resemblances and subordinations among them suggesting [that they are populated by] sentient beings."

Typical of the Victorian era, when science was beginning to have a major say in the issue, was the following passage from H. G. Wells' novel *War of the Worlds*:

> With infinite complacency men went to and fro over this globe about their little affairs, serene in their assurance of their empire over matter…No one gave a thought to the older worlds of space as sources of human danger, or thought of them only to dismiss the idea of life upon them as impossible or improbable. It is curious to recall some of the mental habits of those departed days.

We all know how that story turned out. Wells' account, while fictional, is a response to some of the skepticism within his readership. He must make the case for the existence of "impossible or improbable" extraterrestrial life before going on to describe its nature and motivations. Wells was not alone. Many Victorians doubted that our Earth could be the only planet with life, arguing that there must be someone out among all those stars, and that there must be sibling Earths of distant suns. And what of those nearby worlds, the ones in our own Solar System? Were "intelligences greater than man's and yet as mortal as his own" (as Wells put it) awaiting our phone call?

Ironically, science only added to that yearning by showing us just how large the universe really is. Careful observation and advances in technology demonstrated that the Milky Way was not the whole show but was one galaxy of many in a universe far vaster than anyone had thought.

The Nebular Theory of Planetary Formation

In the late 1700s, two brilliant minds simultaneously came up with the same theory. The German philosopher Emmanuel Kant and the French mathematician Marquis de Laplace independently developed the nebular theory of planetary formation. They hypothesized that our Sun and planets formed out of a great disk of gas and dust.[3] Kant and Laplace posited that the farthest

---

[2] Chalmers, T. (1817) *A Series of Discourses on the Christian Revelation Viewed in Connection with the Modern Astronomy*, Edinburgh, pp. 3–4.

[3] As it turns out, they were right. The Hubble Space Telescope has imaged star- and planet-forming clouds in action.

planets from the Sun condensed the earliest, and so were the oldest. This meant that any civilization on Mars had been around longer than civilizations on Earth or Venus. Sky watchers considered Venus to be a primordial version of our own world, perhaps harboring carboniferous swamps and bubbling carbonated seas. They thought Mars to be a likely abode of life more advanced and ancient in nature than our terrestrial civilization. Astronomers focused on contacting Martians, whose world seemed to evidence telltale signs of massive construction projects, like globe-spanning canals. And while early astronomers were entertaining the idea that nearby worlds might be inhabited, inventors were envisioning how we might get in touch with them (see Chap. 2).

With the advent of modern astronomy and space exploration, we have come to realize that our Solar System has several worlds that could be Earthlike with very little change. If our Sun were just a few degrees cooler, Venus might be an oceanic Eden. As it is, its oceans boiled away early in its history, and its slow spin, coupled with its proximity to Sol, doomed Venus to become our Solar System's hottest planet. Its' surface is a hellish place of searing, intense heat. Mars, on the other hand, is slightly too cold to support liquid water on the surface today. But if it were slightly closer to the Sun, and if it had a magnetic field generated by a slightly larger core, the planet would have water, more atmosphere, and a similar climate to our own.[4] This concept has been understood for quite some time. In 1853, William Whewell formulated his views into a systematic work, penning the essay "On the Plurality of Worlds" (see Chap. 2). In it, he talked about Earth's placement at a perfect distance from the Sun, "which, scorching the inner planets, and driving the vapors to the outer orbs, would make the region of Earth the only habitable part of the system." Whewell's writings were not far from the truth: it only takes a few degrees of temperature to kill a potential Earth, as both Mars and Venus attest.

The Spiral Nebulae: Small and Close or Large and Distant?

As the twentieth century opened, observers could not discriminate between nebulae—those elegant stellar nurseries—and galaxies, great collections of millions of stars. Some of the nebulae, called "spiral nebulae," seemed to have a consistent structure. At the time, astronomers engaged in heated debates about the size of the Milky Way itself. Was it the main part of the universe as

---

[4] In fact, Mars did have an active water cycle, surface lakes and rivers, and perhaps magnetic fields in its infancy.

a whole, with spiral nebulae simply another type of gas cloud like the Crab or the Pleiades? Or were these spirals actually "island universes" like the Milky Way, so remote that their stars blended into an amorphous shape?

These astronomical deliberations of scale culminated in what has become known as "The Great Debate," a discussion between two American astronomers, Harlow Shapley and Heber Curtis. In April of 1920, at the Smithsonian Museum of Natural History, Shapley proved that the Milky Way was larger than believed at the time, and that the Sun's system is not in its center. But Shapley also considered the "spiral nebulae" to be inside the Milky Way, while Curtis argued that they were external to it. Although Curtis was correct, he was assuming the Milky Way's size to be much smaller than it is.

The debate was solved, once and for all, by a new kid on the block named Edwin Hubble. At the eyepiece of the 100-inch telescope of Mount Wilson Observatory northeast of Los Angeles, California, cosmologist Hubble took a series of photographic plates of the "spiral nebula" M31 (now known as the Andromeda Galaxy, see Figure 1.1). His images resolved the spiral shape into thousands of stars. He also spotted several Cepheid variable stars, later used to establish the actual distance to Andromeda.[5] Hubble calculated that the distance to Andromeda was far greater than the size of the Milky Way, proving that Andromeda was an entirely separate galaxy, perhaps similar to our own. An article from the November 3, 1924, edition of the *New York Times* reported that Hubble's recent observations were "striking in their confirmation of the view that these spiral nebulae are distant stellar systems… and that we are observing them from Earth by light which left them in the Pliocene Age." As it turns out, the article greatly underestimated the distances and scales involved.

In ancient times, most people believed the universe was organized around Earth, which was surrounded by an onion-like arrangement of crystal, star-bejeweled spheres. Today we see the universe as a vast wilderness teeming with stars, planets and galaxies. There might be as many Earths out there as there are stars, and there are too many stars to count.

With the modern understanding of the vast cosmic numbers in mind, author Arthur C. Clarke declared, "The idea that we are the only intelligent creatures in a cosmos of a hundred billion galaxies is so preposterous that there are very few astronomers today who would take it seriously. It is safest to assume, therefore, that They are out there…"[6] Wernher von Braun,

---

[5] Because of their stable pulsations in brightness, and a direct ratio between their pulses and their brightness, the luminosity of Cepheid variables can be linked to distance.

[6] *Report on Planet Three & Other Speculations* by Arthur C. Clarke (HarperCollins, 1972)

Fig. 1.1 The "island universe" of Andromeda, a spinning pinwheel of stars 260,000 light-years across. (Galaxy Evolution Explorer image courtesy of NASA)

whose rocketry work led to the first human landings on the Moon, agreed, adding that, "Our Sun is one of 100 billion stars in our galaxy. Our galaxy is one of billions of galaxies populating the universe. It would be the height of presumption to think that we are the only living things in that enormous immensity."[7]

Hello Out There

There's an awful lot of space out there, and it's filled with stars.[8] And as we have pondered the possibilities of other Earths out there, we have gone through a progression, first of wondering if those worlds were inhabited, and then to wondering if we might be able to contact beings there. One of the first serious attempts at communication was made in the twentieth century, at the Martian opposition (close orbital pass) of 1924. The United States

---

[7] Address by von Braun Before the Publishers' Group Meeting in New York, April 29, 1960

[8] Just ask Dave Bowman.

promoted a "National Radio Silence Day" during a 36-h period from August 21–23, with all radios quiet for 5 min on the hour, every hour. At the time, the U. S. Naval Observatory hoisted a radio receiver by dirigible, attempting to hear any Martian broadcasts. The study was uneventful scientifically, if not culturally.

As time went on, the search for extraterrestrial intelligence became marginalized, looked upon with skepticism by the scientific community. But in 1959, all that changed with one paper. Respected physicists Guiseppe Cocconi and Philip Morrison published a paper in the prestigious journal *Nature.* "Searching for Interstellar Communications" outlined the ease with which it might be possible to monitor or transmit radio signals through space, and how those signals might reveal the presence of nearby alien civilizations. Their paper concluded, "If signals are present, the means of detecting them is now at hand…The probability of success is difficult to estimate; but if we never search, the chance of success is zero."

Just a few months after the Morrison/Cocconi article's release, radio astronomer Frank Drake and Cornell astrobiologist Carl Sagan undertook the very first modern radio search, Project Ozma. Both men became outspoken advocates for the search for extraterrestrial civilizations using radio-monitoring systems. Ozma would be the first of many formal attempts to monitor the sky for signals from far-off civilizations, a worldwide search for extraterrestrial intelligence.

Frank Drake felt the odds of discovery were good. His optimism was based on assumptions that he put into the form of an equation, now known as the Drake Equation. Frank Drake was one of the first pioneers to actually try to hear evidence of extraterrestrial life. He also tried to estimate how many planets out there might host sentient beings that were transmitting in radio frequencies. Prior to his work, there had been no formal attempt to estimate the chances of life in the universe at large. Drake devised the equation, in part, as the agenda for a 1961 conference held in Green Bank, West Virginia. The meeting's goal was to discuss the possibility of searching for signals from intelligent alien beings.

Drake's equation attempts to express the variable N, the number of transmitting civilizations in our galactic neighborhood, as the product of seven factors, resulting in the equation

$$N = R * f_p n_e f_l f_i f_t L$$

where:
R_* represents the annual birth rate of stars in our galaxy.

f_p is the subset of stars containing planets.

n_e defines the number of planets per Solar System that are suitable for life (Earthlike worlds).[9]

f_l = the fraction of those planets where life has actually arisen.

f_i is the fraction of planets with life that have evolved intelligent life.

f_t is the fraction of planets with intelligent life that becomes technologically advanced.

L is the average lifetime, in years, of any technological society.

Only the first three elements of the equation are known with any certainty, and the third, the number of Earthlike planets, suffers from wide estimates based on the few that we have located so far. Values for the last four terms in the equation are still completely speculative. Drake himself estimates that N might be as high as 10,000. Carl Sagan, ever the optimist, suggested that transmitting civilizations could number more than a million. Others are less confident, some suggesting that N might only be 1—that Earth possesses the only technologically advanced civilization in existence.

Today, the SETI Institute, in California, continues in the efforts begun by Drake and Sagan. Seth Shostak, the Senior Astronomer and Director of the SETI Research Institute, proposes that despite all the unknown variables, only a couple factors might have the most profound influence in the rise of sentient extraterrestrial civilizations. "The big stumbling blocks are only two: One, do you get life started in the first place? Once you have it started, it's just Darwinian for a while. The second thing is how often do you have a planet with a lot of biology that actually produces thinking brains in the sense that we have one, even if it doesn't think like us? Nobody knows the answer to that."

Homes for ET

Advances in ground-based and space observatories have brought a new understanding to the study of exoplanets, or planets that orbit other stars. Orbiting telescopes such as the Hubble Space Telescope and the Kepler planet-hunter add to our list of known worlds almost daily. It now appears that the majority of stars play host to planets of their own (Drake's f_p variable), and among these we may find hundreds, if not millions, of planets similar to Earth.

---

[9] This number is higher if we include ice moons with oceans, or planets distant from their suns but warmed by tidal friction or large, radioactive cores.

Nevertheless, the ancients may have been right. It may be that our planet simply "lucked out," arising in the right place at the right time. Earth may have won the cosmic lottery when it came to its star, its location in the Solar System, its mineralogical makeup, its status as a planet protected from massive impacts by Jupiter's size and placement. The list goes on. Even our position in the galaxy is of interest. Over 95 % of the stars in the Milky Way may not be able to support habitable planets because their galactic orbits among the stars carry them through the deadly spiral arms of our pinwheel galaxy. The trains of stars that lend structure to our island universe are packed closely together. Any star that circles the Milky Way within one of these glowing arms, and any star that drifts into and out of these arms, is subject to deadly radiation from closely packed surrounding stars. Not so Earth, whose orbit is fairly circular and in sync with the rotation of the rest of the galaxy, keeping it in the more rural space between the spiral arms. Our location may explain why, with so many Earth-similar planets out there, no one has "come to call" in an obvious and overt way.

Drake's radio approach assumes that interstellar travel is far more difficult than long-distance communication using radio waves. And while studies in the 1970s demonstrated reasonable propulsion strategies for getting to other star systems, weakening Drake's primary argument, the search for extraterrestrial intelligence (SETI) is still healthy and alive using many of the world's major radio antennas.

Italian physicist Enrico Fermi worked at Los Alamos in the 1950s and designed the world's first nuclear reactor. He reasoned that if the Sun is a fairly typical star, and there are billions of stars like it in our galaxy, many much older, odds are that there are many stars that host Earth-like planets. If our own world is fairly typical, some of those millions of Earth-like worlds should have birthed life, and among these myriad life forms, many must be intelligent. At least some of those should have developed interstellar travel, something Earth's scientists are considering as you read these words. Even at a leisurely pace, considering the age of the galaxy and the plethora of Earth-like planets out there, the Milky Way Galaxy would have been completely traversed by early civilizations within a few million years. So Fermi wondered aloud to his colleagues over lunch, "Where is everybody?" We'll explore these ideas in depth throughout the next chapters.

Someone to Talk To

Our search for Earths of distant suns has deep roots, and they may be embedded in any or all of these sources, from the biological imperative to pure curiosity to simple survival. Whether our fascination in beings among distant

Earths is a Darwinian attribute or something more deeply emotional or spiritual, humans are driven to search the heavens for company in the great starry night. We now know something about the distant Earths out there, and we are developing techniques to find out what they are like and even whether they host living biomes and beings. With increasing possibilities out there, the search is now upon us in earnest.

2

Early Ideas and Lessons from Our Own Backyard

Moving Off the Center

The ancients of many cultures understood the concept of "world" in the sense that Aristotle did—our Earth held court at the center, attended by the Moon, the Sun, and the planets embedded in their crystalline spheres. Beyond us shimmered the sphere of the distant stars. Alexandrian astronomer Claudius Ptolemy later adopted and refined the Aristotelian, Earth-centered model. His work dominated western thought for the next fifteen centuries.

Many ancient thinkers advocated the concept of the "plurality of worlds." The plurality of worlds concept suggested that many such universes existed, each sovereign and self-sufficient, with an inhabited Earth at the center of each. They based their interpretation on the presupposition that the universe was infinite. Their perspective was also informed by the principle of plenitude, which states that the physical universe must encompass all possible forms of existence. Metrodorus of Lampsacus, a disciple of Epicurus,[1] said, "To consider Earth as the only populated world in infinite space is as absurd as to assert that in an entire field sown with millet only one grain will grow." Others were not so sure. Both Aristotle and Plato opposed the concept of plurality of worlds. Aristotle's concept of the universe described only one Earth at the center of a finite universe.

Sixth century B.C. Greek philosopher Thales of Miletus may have been the first person to wonder aloud about what the universe was made out of.

---

[1] Most famous for his founding of Epicureanism (~307 BC), which taught that pleasure and tranquility were the greatest good.

© Springer International Publishing Switzerland 2017
M. Carroll, *Earths of Distant Suns*,
DOI 10.1007/978-3-319-43964-8_2

He suggested that water comprised most things in the universe, and that Earth floated on a vast sea. Significantly, Thales understood that whatever the universe was made of had a bearing on the possibility of life "out there." His student, Anaximander, took it a step farther, envisioning a universe filled with other worlds arising and dying from an eternal "ether." Although he did not make any specific predictions of life in other parts of the universe, Anaximander did suppose that other Earths—with other beings upon them— might exist.

Medieval thinkers such as Augustine (fifth century) used theological reasoning to champion Aristotle's model. Augustine rejected the idea of a plurality of worlds. He argued that the Incarnation of Christ, designed specifically for the human race, implied that there could be no other inhabited Earths. Other theologians disagreed with Augustine on both logical and theological bases. Albertus Magnus reasoned that since God is omnipotent, he must have created many other worlds. Giordano Bruno also promoted the plurality of worlds' idea, but his motives were not pure. He used his arguments to weaken the Church's position on the uniqueness of Christ. Sadly for him, the Inquisition was catching on at the time, and Bruno is often portrayed as the first person to be martyred in the name of science. A careful reading of history actually contradicts this.[2] Bruno died for his heretical views on salvation, not for his thoughts about the planets.

The sixteenth century ushered in our modern view of plurality of worlds on the heels of Nicolas Copernicus. Copernicus developed the heliocentric— Sun-centered—Solar System. In concert with Galileo's observations of moons moving around planets,[3] astronomers tumbled to the fact that Earth was not the center of all things. The profound implication was that if our world revolved around the Sun, and the distant stars were all Suns like ours, there might be countless Earths revolving around them, a plethora of Earthlike worlds scattered across the cosmos.

Just how distant those worlds were came into focus in the eighteenth century. Astronomer Friedrich Bessel was the first person to suspect that interstellar expanses were far greater than once supposed. In 1838, he was able to measure the distance to the faint double orange dwarf stars of 61 Cygni.[4] He ascertained the distance at 10 light-years, or 200 million times the distance to the Moon.

---

[2] Claims by Carl Sagan and others notwithstanding, Bruno was burned at the stake for his position on several important Church doctrines, along with his views on the Church's politics and theology. Science was the least of his troubles.

[3] In specific, Galileo discovered the four Galilean satellites moving around Jupiter in 1610.

[4] James Bradley first noticed that 61 Cygni was actually two stars in 1753, but the true double nature of the stars, in orbit around each other, was not confirmed until 1934.

Because of its relatively swift motion against the stars, Bessel knew it was one of the closer stars. The neighborhood was getting a lot bigger.

Still, observers began to realize that stars are like our own Sun. This revelation buttressed the assessment that countless Earths could exist in the universe. Whether those worlds were inhabited was another argument altogether.

Although the plurality of worlds was popular in the nineteenth century, it was not universally embraced. William Whewell, scientific theologian and Master of Trinity College, Cambridge, was the first to counter the concept of plurality of worlds using contemporary science. He explored both sides of the argument. In 1853, he pointed out that conditions on other planets in the Solar System are so different from those on terra firma that no known life could ever arise there. At the time, there was also no proof that planets orbited other stars. Whewell also asserted that during most of Earth's history, our world lacked any intelligent beings. But despite his own objections to the idea of life on other worlds, Whewell admitted that "…No one can resist the temptation to conjecture, that these globes …are, like ours, occupied with organization, life, intelligence."[5] As for the stars, Whewell said that "they may have planets revolving round them; and these may, like our planet, be seats of vegetable and animal and rational life…but however many, however varied, they are still but so many provinces in the same empire, subject to common rules, governed by a common power."

Alfred R. Wallace, the British naturalist who cofounded the theory of evolution, argued against the idea of intelligent life in the universe. In the 1905 edition of his book *Man's Place in Nature*, Wallace observed that mankind is the result of a sequence of unique and unpredictable events in the long evolutionary chain. He estimated that the probability that this same chain of events should occur elsewhere—even under Earthlike conditions—is practically nonexistent.

Wallace's reasoning, adopted by some modern biologists, is that a sequence of events may be of little importance when they are taken separately, but in combination, their effects are magnified to such a degree that the final result becomes completely unpredictable. Life, intelligent or otherwise, is not a given on Earthlike worlds. Virtually identical initial conditions can lead to completely different results.

But others reasoned that if Earth was simply a member of a family of planets, then it was likely that even our nearby worlds might share characteristics of our own planet, including seasons, mountains and seas, plantlife, creatures, and perhaps sentient beings. Early Darwinian evolutionary theory held that

---

[5] William Whewell, *Of the Plurality of Worlds*, 1853, London, p. 206.

given time, life should become advanced in intelligence and technology. Mars, long thought to be more ancient than Earth, held the most promise for hosting advanced civilizations, and it was close enough that we just might be able to communicate. But how could we go about it?

Getting the Message Across

Nineteenth century French poet and inventor Charles Cros had heard observers' reports of pinpoints of light visible on both Mars and Venus. Cros assumed that the phenomena heralded the existence of Martian metropolises. He attempted to get funding from the French government to build a giant mirror with which to signal the inhabitants of the Red Planet. His plan was to focus sunlight from the mirror onto the Martian plains, burning diagrams and various shapes into the Martian landscape as evidence to them of our presence across the void. The massive etchings might have lacked the desired effect. Who knows what the poor Martian farmers might think of a brilliant death-ray cauterizing their fields?

Johann Carl Friedrich Gauss—perhaps the most respected mathematician of the nineteenth century[6]—suggested that a giant triangle and three squares, a shape known as "the Pythagoras," could be etched across the Siberian tundra. Ten-mile-wide strips of Siberian pine forest could be left at the center of clear-cut regions, leaving great geometric paths of trees. Within those green lines, the interiors would be cleared and filled with rye or wheat. The hope was that alien beings would see the patterns and recognize that there was intelligent life on Earth.

Joseph Johann Littrow, director of the Vienna Observatory, also advocated the idea of communicating with nearby planets. It may have been Littrow who originally proposed using the Sahara Desert as a gigantic artist's pallet. The idea was to excavate gigantic trenches several hundred yards wide. The canals would transcribe perfect circles, right triangles, and other geometric forms. After filling them with water, engineers would fill the trenches with enough kerosene (called paraffin in those days) to form a skin on top of the water. When lit, the vast shapes could burn as a great beacon throughout an entire night. Different shapes could be illuminated on consecutive nights, showing the inhabitants of nearby planets that we are here, and we know something about geometry, at least (Fig. 2.1).

---

[6] Gauss's mathematics led to the rediscovery of the first asteroid, Ceres; the object disappeared behind the Sun soon after Piazzi discovered it, but Gauss successfully calculated where it should be.

Fig. 2.1 Proposed techniques for contacting Martian civilizations included incandescent, kerosene-filled canals in the Sahara Desert (*left*) and geometric shapes made of Siberian forests and planted wheat. (Art © Michael Carroll)

The prospect of making contact with extraterrestrial beings was all the nineteenth-century rage. In 1891, the Prix Pierre Guzman was established. A bequest from the estate of Anne Emilie Clara Goguet, the award was named after Anne Emilie's son, a major in the French army. Two prizes were to be given, one for medicine and one for astronomy. In the case of the astronomical one, a prize of 100,000 francs was up for grabs to the first person who succeeded in communicating with another planet. According to her will, the award was to be granted to anyone "who shall discover a way to correspond with one of the heavenly bodies—that is to say, to receive an answer from the inhabitants of a planet to some sign made to them on ours." Mars was pointedly excluded from the contest, as it was considered too easy a target. If no communication was made, the accumulated interest over each successive five-year period would be awarded to those making significant contributions to the field of astronomy.[7]

The twentieth century opened with the invention of radio transmitters and receivers, devices that would revolutionize global communication. Italian Guglielmo Marconi is generally credited with bringing practical radio into the public realm. In 1901, Marconi successfully transmitted a radio signal across the Atlantic Ocean, from Cornwall, England, to St. Johns, Newfoundland, proving that radio signals could communicate over long distances. In the early

---

[7] Prizes were awarded to French astronomers Louis Fabry and Henri Joseph Perrotin in 1905, and to Maurice Loewy, Viennese-born director of the Paris Observatory.

Fig. 2.2 Marconi, at *left*, supervises the raising of a communications kite at St. Johns, Newfoundland. (Courtesy of Wikipedia commons: https://commons.wiki-media.org/wiki/File:Marconi_at_newfoundland.jpg)

1920s, Marconi claimed to have picked up signals from an interplanetary source (Fig. 2.2).

Marconi had company in the United States. When he wasn't investigating electrical currents or radio waves, inventor Nikola Tesla used his fertile mind to turn toward the problem of interplanetary communication, too. As early as 1893, Tesla lectured on the possibility of radio transmission as a mode of communicating over great distances. Five years later, Tesla demonstrated a remote-controlled boat, called a "teleautomaton." He asserted that his inventions might be used to communicate with other worlds. Tesla erected a 6-m-tall tower strung with copper wiring to monitor radio signals. In 1901, the inventor wrote that his tower was picking up radio "disturbances" that he attributed to interplanetary communiqués. Tesla wrote that "The feeling is constantly growing on me that I had been the first to hear the greeting of one planet to another."

In the March 20, 1920, issue of *Scientific American,* H. W. Nieman and C. Wells penned the article "What Shall We Say To Mars? A System for Opening Communication Despite the Absence of any Common Basis of Language." The authors proposed a series of dots and dashes, resembling Morse code, along with simple pictures, to communicate with beings on other planets in the Solar System. Their approach foreshadowed later efforts using messages bolted to the side of spacecraft (see Chap. 7).

The year 1924 saw an opposition of Mars and Earth, a point at which both planets passed nearest each other in their orbits. The U. S. Army combined forces with Marconi and others to listen for Martian signals. While many noted radio signals, there was no evidence of extraterrestrial sources. In fact, the frequencies used in those days were largely deaf to signals outside Earth's atmosphere. Their low frequencies caused signals to bounce back from the ionosphere. (In fact, the ionosphere made Marconi's transatlantic communication possible, with radio signals from England reflected off the ionosphere to radiate over the horizon, making their way to Newfoundland.) Only later did experimenters come up with equipment able to receive signals of higher frequency, the kind that could travel between stars.

Several researchers claimed to have received transmissions from cosmic sources. Competition was beginning to grow. In 1937, Nikola Tesla petitioned the Guzman overseers to award him the prize for his—as he put it—"discovery relating to the interstellar transmission of energy." He was not a recipient.

Advances in radio equipment during and shortly after World War II made technology readily available. The new radio antennae could study the skies in higher frequencies that penetrated Earth's atmosphere. The time was ripe for the modern search for extraterrestrial signals coming from distant Earths. But where would we look?

The Goldilocks Principle

As astronomers gained further insight into the nature of the Solar System, it became obvious that Mars was too cold—and its air too thin—to support liquid water on its surface. This is significant, says NASA Ames exobiologist Chris McKay. "On Earth, wherever you have life, you have some access to liquid water, even if that means water vapor. In places on Earth where the water is frozen, like on the Greenland ice cap, it's essentially lifeless and sterile. But even at those margins where the ice is melting, you get microbial mats and other biology. Life on Earth requires liquid water."

If liquid water is a necessity of life, Mars was looking more and more dismal as an abode for it. And on the other side of Earth, cosmic neighbor Venus was far too hot, simmering at some 462 ° C (864 ° F) on the surface. Venus receives nearly twice the solar energy that Earth does, while Mars' portion is a dim 50 % of that of our own world. Like the fabled porridge of Goldilocks, Earth was not too hot and not too cold, but just right. The idea led scientists to realize that every star has a zone in which liquid water could, theoretically, exist on the surface of a planet (Fig. 2.3). These zones have come to be called "habitable zones."

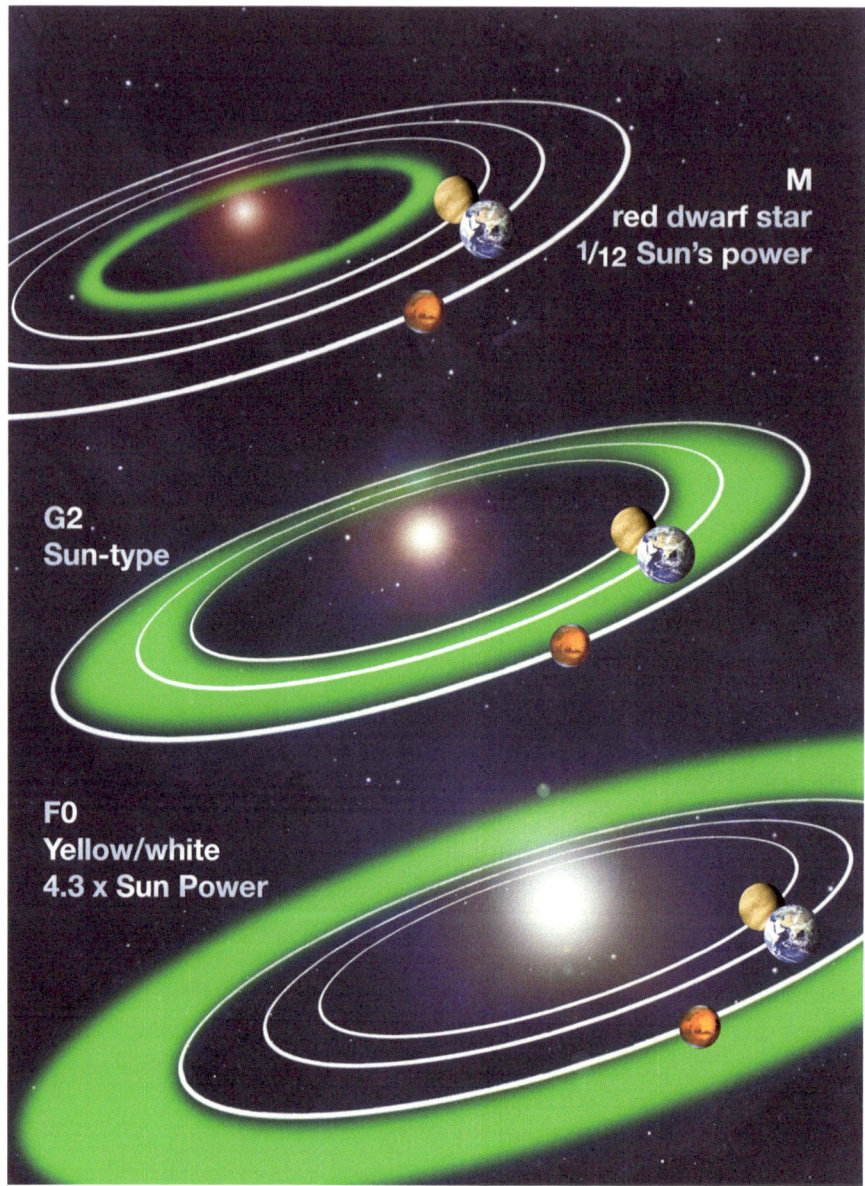

Fig. 2.3 The habitable zone of an exoplanet is determined by the size and temperature of its parent star. Here, our familiar paths of Venus, Earth and Mars are superimposed on several example systems. (Art © Michael Carroll)

The habitable zone of a star varies with the nature of the star itself. The habitable zone of a hot young sun will arc around the star at a greater distance than the habitable zone of a cool red dwarf. An exoplanet (a planet beyond our own solar family) needs to be farther away from a hot star for conditions to have liquid water. An M star (sometimes called a red dwarf) is not only cool but also smaller. The surface temperature hovers around 3000 °C (the Sun reaches 15 million degrees Celsius), glowing like an ember just three thousandths as bright as our Sun. Nearly 80 % of all stars in the universe are M stars. The habitable zone of a small star is close to itself, where its feeble heat can still allow liquid surface water.

Migrating Habitable Zones

The shift in a habitable zone was a concept first understood by astrophysicist Michael Hart. In 1978, Hart calculated that the Sun's habitable zone began farther in, and has migrated out over time as the Sun has aged and grown brighter. Four billion years ago, during Earth's formative years, Sol was only 70 % as bright as it is today, and Sol continues on this path, forcing our habitable zone outward. His analysis showed that Earth is in a central-enough location that it will remain in the habitable zone throughout this shift, but just barely. Hart's research demonstrated that had Earth formed only 5 % closer to the Sun, our planet would have experienced a runaway greenhouse effect, making us more of a twin to Venus than we already are. Had Earth coalesced 1 % farther away, the home world would have suffered a runaway glaciation, with the surface and oceans freezing over permanently. Hart's models portrayed both of these situations—chilled or cooked—as irreversible.

Hart's work led to a new term in exoplanet research: the continuously habitable zone. This is the sweet spot within a habitable zone where conditions remain stable for liquid water, despite the shifting of the habitable zone around it. The continuously habitable zone provides yet another limiting factor in our search for Earth's exoplanet twin. If Earth's orbit were a bit more out-of-round than it is, its path would also take it out of the continuously habitable zone as time progressed. But another implication is that a frozen planet on the outer edge of the zone may become habitable as its star ages and warms. Habitable worlds closer in may become too hot to support life as their suns brighten.

The continuously habitable zone in other star systems may be quite different from that of terra firma. Sol's continuously habitable zone is fairly thin, but for smaller suns like M stars, it may not exist at all. Still, stable regions within habitable zones of cooler stars, whose lifetimes are drawn out, last longer than those for short-lived suns because the habitable zones migrate over a longer period of time.

Hart's estimates are now thought to be too narrow because of several factors that have recently come to light. Climate stabilizers appear to be built in to the design of Earth. The most influential of these is a chemical recycling process called the CO_2-rock cycle. Carbon dioxide (CO_2) is a powerful greenhouse gas. Without it, Earth's surface would average 40 ° C colder than it does, because the CO_2 helps to hold the heat in. In a complex feedback loop, the CO_2-rock cycle stabilizes global temperatures. As Earth warms, increased weathering removes CO_2 from the air. That reduction in the greenhouse gas leads to a global drop in temperature. As the temperature sinks, the weathering processes abate, allowing the CO_2 levels to rise again.

A second stabilizing process—subtly different—depends on plate tectonics. The rocky surface of Earth acts as a chemical sponge, soaking up various gases and chemically locking them into the rock. (Mars is red because its oxygen has been locked away into the surface, "rusting" the planet.) Earth's plate tectonics bring the gas-impregnated rock down into the mantle, where it melts, freeing the chemicals and minerals. Then, these materials are recycled back into the environment through mountain uplift or volcanism.

Taking into consideration the CO_2-rock cycle and other moderating factors, atmospheric expert James Kasting has revisited the concept of the continuously habitable zone. Kasting estimates that the Sun's continuously habitable zone spans a distance of 0.95 AU to 1.15 AU from the Sun. Although this is more optimistic than Hart's estimate, it is still a thin portion of the current habitable zone. Earthlike planets with surface conditions that remain habitable over long periods may be rare indeed.

To understand habitable zones, those regions around distant suns where we might find Earthlike worlds, we must first take a look at the grand menagerie of stars in the cosmos.

Starring Roles

The most familiar star to any life form on Earth is the Sun. Sol dwarfs every other body in the Solar System, weighing in at a mass equivalent of 1,300,000 Earths. Our nearest star grips our entire planetary system—terrestrial and giant planets, asteroids and comets—in its powerful sway. Its heat drives our wind and ocean currents. It feeds energy into our biome, where it is converted into plant sugars, and then moves on through the food chain.

As stars go, the Sun is steady and sedate. Its surface simmers at 5500 °C, while temperatures at its core reach more than 15.5 million degrees Celsius. It radiates a steady flow of charged particles known as the solar wind. This stream moves at about 450 km a second throughout the Solar System. Occasionally, a solar flare erupts, sending a barrage of particles throughout the Solar System. Flares can interrupt satellite communications or knock out power grids, but these events are tame compared to some of the stars we will visit. The stability of the Sun may well have contributed to the extent and duration of life's rule on our planet.

Just as Kant and Laplace suggested, the life cycle of a star begins as a vast disk of gas. The universe is filled with elegant, glowing gas clouds called nebulae, and many of these clouds are the birthplaces of stars. Hydrogen is the most common element in the universe, so it makes sense that nebular gas is mostly hydrogen. Some nebulae incorporate the detritus of exploded stars. These nebulae also contain heavier elements such as metals, and these are the nebulae that will lead to an Earthlike world.

As drifting gas condenses into dense knots and tendrils, it begins to develop gravity. The more dense the gas, the more gravity it has. But the motion of gas moving within a cloud follows certain patterns dictated by physics, including eventually spiraling toward the center. The spinning gas flattens out into a disk. Its central bulge, where most of the mass is, pulls gas and dust in radially, becoming a globe called a protostar. The central sphere gains mass and gravity, finally collapsing in on itself and breaking atoms down in its core, triggering nuclear fusion. Within the surrounding disk, eddies and currents are also forming dense spots, and these become planets. Telescopes such as Hubble and Spitzer have resolved images of disks around stars, and in some cases have been able to actually see planets forming within those disks.

Once the star begins to burn its hydrogen fuel, it pours out a gale of particles, called the solar wind, across its planetary system. This wind clears out the gases near the star, leaving many terrestrial—or solid-surfaced rocky—planets in its wake. Giant planets form farther out, where gases are calmer and temperatures cooler, although they may not stay there (see "Arranging Planetary Families," below).

Stars spend most of their time, roughly 90 % of their lifespans, burning hydrogen into helium. This very stable time of life is referred to as the "main sequence," where stars burn steadily and brighten over time. The energy of the hydrogen fusion "holds up" the star's outer layers above its core.

Stars are classified into seven categories, according to their spectrum and from hottest to coolest. The more mass a star has, the brighter it is and the faster it burns its fuel. The first stars, inhabiting the early universe, all consisted of hydrogen and helium. But once they began to explode as novae or supernovae, the scraps of their explosive deaths resulted in heavier elements. Those elements later combined to make new generations of stars. These later-generation stars, our Sun included, still have a majority of hydrogen and helium, but also contain heavier elements such as iron, lithium and calcium. All of these elements contribute to the construction of Earthlike worlds.

Fig. 2.4 A dual planet orbiting Proxima Centauri (*star at right*) might be cool enough, and its conditions just right, to hold atmospheres and even liquid water on the surface, as seen on the distant planet. The primary stars Alpha Centauri A and B are at *left*. Alpha Centauri A, at far *left*, is Sun-like, while its companion is a cooler, K-type star. (Art © Michael Carroll)

The size of a star determines its longevity. The longer a star lives at a stable stage, the more chance Earthlike worlds have of getting started. Large suns use up their fuel quickly, burning bright and hot but not for long. The large star Spica, ten times the mass of Sol, should live "short and fast," lasting about 10 million years. Our Sun will have a lifetime spanning the course of 10 *billion* years. The small, cool dwarf star Proxima Centauri will last perhaps 90 billion years, into the old age of the universe itself (Fig. 2.4).

M Dwarfs

Proxima is among the smallest stars, called M dwarfs. These little suns boast the longest lives. Also known as red dwarfs, they range from 0.075 up to half of Sol's mass. Because of their slow-burning natures, some red dwarfs may live up to 600 billion years. While larger stars collapse after burning through the hydrogen in their cores, red dwarfs burn all of their hydrogen, from top to bottom, gradually. Instead of a deadly helium build-up like their larger siblings, M dwarfs have very energetic internal mixing. This active convection keeps the helium and hydrogen well mixed. Eventually, the helium begins to take over. Then, like their larger cousins, M dwarfs collapse into small white dwarfs.

M dwarfs are the most common stars in the galaxy, says NASA Ames Senior Research Scientist Thomas Barclay. "We have all these classes of stars, and then we put everything at the bottom—which is 70 % of the stars known—in one bin. But M dwarfs range from things smaller than Jupiter—Saturn-sized stars—all the way up to half the size of the Sun, that behave much more similarly to the Sun. If you say you study M dwarfs, it means that you study most stars."

For planet-hunters, red dwarfs present some advantages. Using the radial velocity technique, which detects changes in starlight as a planet moves its star (see Chap. 3), the presence of a small terrestrial planet is more obvious than with larger stars, because the planet's gravity will pull more markedly on the small M star. Since their HZs are closer in, the influence of nearby Earthlike worlds will tend to be more obvious than ones orbiting farther away from their parent stars. Another planet-finding approach, the transit method, also benefits from the diminutive size of red dwarfs. Detecting dips in the light of the star's surface as a planet moves in front, the small face of a red dwarf will be blocked more dramatically than the larger sphere of a Sun-sized star.

K-Type Stars

What about Earthlike worlds orbiting other types of stars? Similar to our Sun are the K-type stars, sometimes referred to as orange dwarfs. Slightly hotter than red dwarfs, these small stars burn at about three-tenths of Sol's brilliance. Their low temperature extends their lifespans, giving them 15 to 30 billion years to live out their existences. They are three to four times as common as Sun-like stars. The smaller of the two main stars in the nearby Alpha Centauri system is a K-type star. Its habitable zone will be similar to the Sun's, but the dynamics of any orbiting planets there will be skewed because of the presence of other stars next to it.

G-Type Stars

Our Sun is a G-type star, a type that makes up 7 % of all main sequence stars, which means that 7 out of every 100 stars has conditions favorable to hosting Earthlike planets. G-type stars range in mass from 0.8 to about 1.2 solar masses. Nearby Alpha Centauri A is a G-type star. Both it and the Sun have life expectancies of about 10 billion years. (Sol has already burned through about half of that time.)

Our Sun is a main sequence star, classified as a G2 yellow dwarf star. Nuclear fusion in its core burns hydrogen, converting it to helium and generating an incredible amount of energy. As a star ages, its hydrogen begins to run out, and what's left is a heavy shell of helium. The star's core begins to compress even more, abandoned by the supportive force of hydrogen fusion. The helium on the outside, along with leftover hydrogen, expands and heats up even further, and the star grows larger and brighter, often transforming into a red giant. During this phase, our Sun will expand to fill the orbits of Mercury and Venus, perhaps even making it out as far as Earth's orbit.

In this violent fuel-burning process, stars the size of Sol fuse heavier elements such as carbon. Larger stars generate more varied elements. Finally, when there is no more fuel for the star to burn, it departs from the "main sequence" and begins to die. Size determines the fate of a star. Medium to low-mass stars such as our Sun swell into a giant star, and eject their shell of spent hydrogen and helium into space. What's left is a "dead" star called a white dwarf. Its habitable zone shrinks down to a narrow band around the star, leaving any Earthlike worlds out in the cold.

A star with four to eight times the mass of the Sun expires in an enormous blast called a supernova. A typical supernova may put out as much power in an instant as an average star does during its entire lifetime. The resulting wave of gas expands at up to 10 % of the speed of light. Any Earthlike worlds nearby would be destroyed outright by the explosion or fried by its radiation.

F-Type Stars

F-type stars are even rarer, accounting for 2 % of the stars in the universe. These stars span up to 1.5 solar diameters, with luminosities seven times as bright and temperatures of 6500 ° C. Burning their fuel hot and fast, these stars are active for only 2 billion years. About 37 F-type stars lie within 50 light years of us. Due to their heat and radiation, the habitable zones of F stars are at a greater distance, but the short life span of F stars may preclude any stable planetary biomes.

A-Type Stars

The A-type stars make up just 1 % of the stellar population. These hot stars burn with a surface temperature of 8000 °C and a brightness of twenty times that of the Sun. Having twice the mass, they end life in about a billion years.

The brightest star in Earth's sky, Sirius A, is one of these. With an even shorter stable period than F stars, finding Earthlike worlds with active biology in these systems is unlikely.

B-Type Stars

The blue giants, classified as B-type stars, are incredibly hot. At a thousand times the luminosity of our Sun, the searing temperatures of B stars reach 15,000 °C. They contain between 2 and 16 solar masses. B-type stars rotate quickly, with equatorial speeds reaching 200 km/s. They generate fierce stellar winds of up to 3000 km/s, a deadly hurricane to any nearby planets. B stars collapse into cold, dead white dwarfs within 50 million years of their birth. Only one in every thousand stars is a B-type. In our search for Earths of distant suns, B stars are not a promising place to look.

O-Type Stars

Most scarce and deadly of all stars is the O-type, known as a blue supergiant. Supergiant surface temperatures surpass 50,000 °C. With up to 90 solar masses, they burn out in half a million years. Under the light of a million suns, the presence of an Earthlike planet is difficult to imagine.

From Supernova to Black Hole

More exotic stars lurk in the void. Most are unlikely to have anything resembling a habitable zone, so any Earth-sized world there would certainly not be Earth-like. Supernovae, exploding stars, contain cores that burn to a point at which they carry out a different type of fusion—called carbon fusion. When the core reaches this stage, the star explodes with a luminosity of 5 billion times its original brightness. Other supernovae are the result of the collision of two white dwarfs. Supernovae can also be triggered by the collapse of a star's core.

The brightest supernova on record was initially detected on June 14, 2015. Before fading again, the star fleetingly reached the brightness of 600 billion Suns. The titanic exploding sun, logged as ASASSN-15lh, was a member of a rare class of supernovae called "superluminous supernovas."

As the core of some supernovae continues to collapse, the entire star—often the mass of the Sun—shrinks to only 20 km in diameter. The core's gravity draws in upon itself, and protons and electrons clash together to make neutrons. This kind of shriveled sun is called a neutron star. The gravity of an average neutron star is 2 billion times as powerful as that of Earth. The explosion from the neutron star's collapse spins the core rapidly—up to 700 rotations per second—spewing out radiation like a lighthouse sends out beams of light. Its formative explosion would clear its inner planets of all atmosphere, and its pulses of radiation, usually X-rays and gamma rays, would leave any nearby worlds cold, barren and soaked in radiation.

If large enough to begin with, the core of a collapsing star will continue shrinking until it becomes a point in quantum space/time. This bizarre stellar phantom is called a black hole. Black holes do not generate the kind of energy that a main sequence star does, and consequently they have no habitable zones, making them a poor target in our search for Earthlike worlds.

This survey of stars shows us that not all stars are candidates for possessing Earthlike planets. Those that *are* retain varied habitable zones, dependent on the nature of their host stars. Stars must remain stable for long periods in order for an Earthlike world to arise. They must not put out too much heat or other radiation, and they must contain elements that can contribute to the building of rocky planets.

Habitable Zone Types

Since the nature and stability of a habitable zone is dependent on its host star, which are the best candidates for finding life? Types O and B stars live such short and furious lives that Earthlike planets don't get the chance to form out of the stars' accretion disks. By the time things settle down for any new terrestrial planets, O and B stars have begun their death throes.

Suns of the type A and F show more promise. With stable lifetimes of between 1 and 2 billion years, planets can form, and biomes may even become established on their surfaces. Life on Earth has left its evidence as far back as 3.8 billion years ago, less than a billion years after star birth. But because of their higher temperatures, these stars have a higher energy output, especially in ultraviolet (UV) radiation. UV radiation can be deadly to life, breaking down the bonds that hold together complex organic molecules. But it will depend on the planets circling them. UV radiation cannot penetrate deeply into soil, ice or water. Some evidence hints that life on Earth began on the ocean floors around volcanoes. Any Earthlike world with lakes, seas or oceans might provide plenty of shelter for microbes and more complex life forms. But the relatively short lifespan of A and F stars means that intelligent life may not have a chance to take hold. Humans did not arrive on the scene until Earth had been around, next to a stable Sun, for roughly 4 billion years.

Some 90 % of all stars belong to the K and M classes. K stars (the orange dwarfs) are slightly smaller than our Sun, weighing in with masses ranging from 0.45 to 0.8 the mass of our own star. They remain on the main sequence for 15–30 billion years, plenty of time for planets to form. In addition to their stability, K stars produce far less UV radiation than Sol does. Although their lower temperature means that their habitable zone is closer to the star, reduced radiation may mean that they are candidates for Earthlike conditions on nearby worlds.

The long life and low temperatures of M stars make them good candidates for finding other Earths. The habitable zone of an M dwarf forms a narrow ring around the dim orb. If our Sun were an M dwarf, its habitable zone would extend from just $1/10$ to ¼ the distance of Mercury. The habitable zone is so close to the star itself that any planets within it are likely to be tidally locked, always keeping one face toward their sun. Temperatures on the sunlit side might soar, while conditions on the night side would chill precipitously.

At one time, it was thought that M stars were starved of the elements that contribute to the construction of terrestrial worlds. Mercury, Venus, Earth and Mars are composed of nickel, silicon, iron, aluminum, calcium, magnesium and assorted other elements. Some early studies of stellar spectra indicated that these elements were rarer in M stars. But the Southwest Research Institute's Dirk Terrell says, "We haven't seen any red dwarfs that lack metals, the ones you'd expect to see at the earliest stages of the evolution of the universe." Research indicates that there are few major differences in elemental abundances between M stars and the Sun.

Recent work provides further hope for finding life in M-star systems. Not all planets in tight orbits need to be tidally locked. Ocean currents, large moons or thick atmospheres can provide inertia to keep a planet rotating. Additionally, even if a world becomes tidally locked so close to its star, computer models demonstrate that the circulation patterns in an atmosphere can play a critical role in subduing the wild temperature extremes from day to night. Simulations showed cloud cover developing on the day side of an Earth-mass world, keeping temperatures down and air pressure high. Winds generated by that high pressure on the lighted side redistributed the heat to the night side, raising those evening temperatures.

We see a similar situation on our own world. A constant rise and fall of currents mixes the atmosphere well in its lowest layer, called the troposphere. Solar heating and currents welling up from the warmed ground keep the air well-mixed, moderating our temperatures. Our weather is simply a mechanism in constant search of equilibrium in our planet's temperatures. On the day side of the globe, hot air forms over the region of the planet pointing most directly at the Sun, near the equator. There, hot air rises, moves away from the equator and drifts toward the poles. It cools at high altitude and sinks back down, migrating at low altitude back toward the equator. This airy conveyor-belt of atmosphere is called a Hadley cell. Planets orbiting red dwarfs may benefit from energetic Hadley cells, enabling life to take hold there.

For Earth-like exoplanets not locked in a deadly tidal dance, another roadblock to life may be the nasty red dwarf habit of flaring up. M dwarfs often send out energetic tongues of material, doubling in brightness as they discharge massive radiation, sterilizing the planetary neighborhood around them. But these flares could also trigger the production of ozone in planetary atmospheres, and this ozone would provide a shield from such radiation.

In addition, what red dwarfs lack in potential habitable real estate, they make up for in sheer numbers. Accounting for nine-tenths of the 100 billion stars in the Milky Way, M dwarfs may play host to 60 million Earthlike worlds.

Still, it seems that a complicated combination of factors contribute to Earth's own habitability. In all the millions of worlds of our galaxy alone, is it possible that Earth is unique?

Past Lives of Our Earth

Earth has not always been the life-nurturing paradise that it is today. Present-day conditions are quite different from what they were when life first arose on the planet. Temperatures and oxygen levels have ebbed and flowed. If we could travel back in time to earlier stages of our Solar System, we might not recognize Earth at all.

Earth's progression of development may well be representative of some of the Earthlike planets we will discover in other star systems. In the formative stages of our own Solar System, planets formed within the great primordial cloud from which the Sun also came. Earth was cocooned in a "reducing" atmosphere rich in hydrogen. In a phase known as T-tauri, the young Sun's solar wind cleared out the dust from the inner Solar System. It also stripped away much of our planet's first atmosphere. Comets and asteroids fell onto Earth and its fellow planets, carving out big craters and leaving behind water and minerals. Early in the process, a Mars-sized space rock slammed into the edge of Earth, creating a ring of debris that became the Moon (see "Isn't that special?" below).

Earth's first oceans appeared, but they were not oceans of water. Instead, the face of our world was awash with seas of molten rock. The great fusillade of asteroids and comets continued to rain destruction upon the glowing landscape. Their deadly salvo continued in full force until about 3.8 billion years ago. The infant Earth was a scant 4 or 5 million years old. Even then, the planet was differentiated: heavy materials had settled to the center, with lighter stuff migrating up to form a crust. Radiogenic materials within the core, along with the leftover heat from its creation, raised internal temperatures, triggering worldwide volcanic activity. These erupting mountains, in concert with materials from impacting comets and asteroids, replaced the dwindling reducing atmosphere with new gases. Carbon dioxide, nitrogen and water vapor filled the skies as Earth's second atmosphere. As they did, clouds condensed, and the first true rains fell onto cooling lava rock (Fig. 2.5).

This Dantean version of Earth gave way to another quite alien world. As the first watery oceans rolled across Earth's surface, the only upraised dry land areas were the rims of craters. But eventually, a new world-building force came into play, that of plate tectonics. Earth is like a jigsaw puzzle. Continents float on rock rafts called plates. These plates move and bump into each other on a "sea" of soft hot rock beneath Earth's crust, the mantle. As the plates ram into

Fig. 2.5   A primordial Earth peers down on a Moon still volcanic from the fires of its creation. Landmasses on Earth are limited to upraised crater rims; plate tectonics are still millions of years in the future. (Art © Michael Carroll)

each other, rock gets pushed up into mountains or melted underneath in a process called subduction. Earth's rocks soak up the gases in our atmosphere, chemically binding carbon dioxide and other gases to the surface. But the rocks eventually melt when plates collide. The trapped gases are spewed back out as volcanoes replenish our atmosphere.

This oceanic stage of Earth presented vast stretches of water where new continents rose. The landscapes of our world were barren and sterile. Craters continued to form though at a much-reduced rate, and some of the largest asteroids left scars that we still see today. South Africa's Vredefort dome, Australia's Warburton impact basins and the Yucatan's Chicxulub crater are the remnants of ancient impacts.

How hot were the ancient oceans? Initial estimates put global seas at nearly the boiling point 3.5 billion years ago, but new research indicates that this may not have been the case. Far from being a simmering inferno, early terrestrial conditions may have been downright chilly. Studies[8] of the isotope oxygen-18 in samples from South Africa's Barberton Greenstone Belt give biologists new insight into early biomes. The site contains some of the most ancient preserved rocks on Earth. Those isotopic studies have revealed that the rocks formed under cool conditions far below the boiling point of water. Additionally, researchers found the presence of clay-like diamictite, a sediment usually formed in glacial environs. Adding to the picture is 3.5 billion-year-old gypsum, which would have formed in deep, cold seawater. The samples also displayed varved sediments—seasonal bands typically laid down when standing bodies of water freeze. Even magnetic data back up the chilled scenario. Paleomagnetic data helps to lock down the location at which the rocks initially formed. The data shows that these samples formed near tropical latitudes.

Because even these near-equatorial rocks were cool, temperatures farther north or south must have been even cooler. Researchers suggest that earlier estimates of higher temperatures sampled brief periods of increased hydrothermal activity, but the new work takes into account ambient overall temperatures rather than localized ones. If the new estimates are accurate, our planet has hosted conditions amenable to life for far longer than originally thought.

Earth continued to change and shift. Continents rearranged themselves, prompting changes in sea currents and weather patterns. Craters came and went as weather and mountain-building eroded the terrain. Minerals from highlands washed into the ocean basins, where they were recycled back to the continents by the subduction and uplift of plates. But on the microscopic

---

[8] See *3.5-Ga hydrothermal fields and diamictites in the Barberton greenstone Belt* by Maarten J. de Wit and Harald Furnes; *Science Advances*, February 26, 2016.

level, something remarkable was taking place, something that would eventually transform our alien Earth into the world we inhabit today.

Within puddles and tidal pools at the ocean's edge, life began. How did it happen? What do we find in the chemistry of the rocks and in the fossil record? Life's origin is a rich area of research for biologists today, and a topic for theologians and philosophers as well. One thing is certain: over 3.8 billion years ago, Earth's chorus line of life took to the stage. Single-celled microbes appear as fossils from about that time, but these little creatures are complex, with delineated internal structures and varied outer membranes. Apparently, life had been around for some time before these fossils were put down. Microbial life transformed the atmosphere of Earth, pumping O_2 into the air. Many researchers consider this a third atmosphere of Earth, one that enabled life forms with oxygen-friendly metabolisms to thrive.

At some point, life came ashore, but the world was a changeable, dangerous place. Earth's toxic atmosphere was laced with lightning and furious winds. Meteors continued to fall. Eventually, new life forms such as the plankton and plants brought an influx of oxygen into the air. Global temperatures varied dramatically. During the age of the dinosaurs, high temperatures may have prevented any extensive ice—even at the poles—but the planet has also undergone radical temperature swings into the cold. Some geologic and isotopic studies[9] indicate that our world may have suffered a planet-wide deep freeze. Temperatures fell to well below 0 °C across the map, resulting in a complete glaciation from pole to pole. The world's oceans would have frozen over in episodes that lasted approximately 100 million years. These events are referred to as Snowball Earth ice ages. How Earth pulled out of such an environmental catastrophe is not clear, but the timing of Snowball Earth glaciations seems to coincide with several mass extinctions, along with the advent of oxygenating organisms long before life came on land.

Between the planet's ice ages, ocean temperatures were actually substantially higher than they are today, with ocean levels some 200 m higher than we currently see. Global temperatures may have clocked in at 8 to 10° warmer than today, which probably resulted in higher humidity. Further back, between 3 and 3.5 billion years ago, ocean temperatures may have topped 85 °C, deadly by today's standards. The fossil record shows us that during that time, life inhabited the oceans. These "unearthly" swings in temperatures and gasses, combined with the extreme biomes in which life thrives here, may indicate that the current environment of Earth is not, in fact, the best model to use in the search for extraterrestrial life.

---

[9] Hoffman and Schrag, 2000.

The Great Dyings

Many of the planet's species disappeared during the Permian extinction, the largest extinction event yet detected. Life on our planet did not arise in a steady progression. It accelerated and diversified, only to be beaten back again or to stabilize at a balance. Earth has been subjected to several mass extinction events. Topping the list, the Permian extinction may have wiped out 90 to 96 % of all species on land and in the oceans.[10] The Cretaceous-Paleogene (or K-T) event is the most famous, as it ended the reign of the dinosaurs. But life on this planet was nearly wiped out multiple times (see Fig. 2.6). What this means in our search for other Earths is that although we may find sister planets, any past life there may have been snuffed out in a Permian-event-on-steroids. The same fate could very probably befall Earth. Many see this as an impetus for humanity to spread to other worlds. In fact, astronomer Carl Sagan saw this fact as one reason that advanced civilizations must travel among the stars. In his book *Pale Blue Dot*, Sagan asserted, "Since, in the long run, every planetary society will be endangered by impacts from space, every surviving civilization is obliged to become spacefaring—not because of exploratory or romantic zeal, but for the most practical reason imaginable: staying alive."

The Earth's roller-coaster story of planetary evolution and life enables us to imagine what various forms Earths of distant suns may take. But the cosmos has provided us with other insights very close to home.

Earthlike Planets in Our Solar System

One way to gain understanding about potential distant exo-Earths (Earthlike exoplanets) is by studying the worlds near us. With Earth as the primary example for imagining habitable planets in other star systems, early astronomers struggled to envision what other Earths might look like. They had two other specimens of somewhat Earthlike worlds just next door. Earth is bracketed by two planets that can be considered somewhat "Earthlike," Venus and Mars. Even small Mercury has a similar internal structure. Each of these three terrestrial, or Earth-similar, planets has an outer crust overlaying a lithosphere (or upper shifting mantle), a deeper silicate mantle and a metallic core.

---

[10] Recent controversial research suggests that the Permian die-off was not as widespread as once thought. Land species believed to have gone extinct during the event have been found later in the fossil record. For more, see Gastaldo, et al. "Is the vertebrate-defined Permian-Triassic boundary in the Karoo Basin, South Africa, the terrestrial expression of the end-Permian marine event?" *Geology*, October 2015.

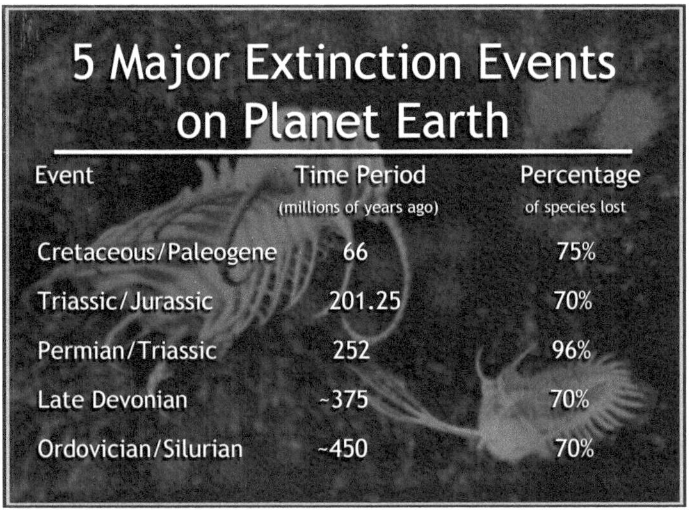

5 Major Extinction Events on Planet Earth

Event	Time Period (millions of years ago)	Percentage of species lost
Cretaceous/Paleogene	66	75%
Triassic/Jurassic	201.25	70%
Permian/Triassic	252	96%
Late Devonian	~375	70%
Ordovician/Silurian	~450	70%

Fig. 2.6 Planet Earth has endured many major extinctions throughout its history, as shown in this table. Exoplanets undoubtedly suffer the same events. How do these influence the likelihood of life on other worlds?

As for planetary interiors, we know the most about Earth's. We know what the surface is made of, and we've had tastes from the deep interior by way of lava eruptions and deep drilling projects. The planet gives us an added bonus. Seismic activity (volcanic eruptions, earthquakes) sends shockwaves through the planet's subsurface. Because we have deployed seismometers around the globe to monitor such waves, we can chart the subtle movement and arrangement of the interior. Like a bat using echolocation in the dark, the way seismic waves bounce around inside Earth reveals its structure. Just 4/1000th of our planet's mass resides in its crust, a hide that varies in thickness from about 6 km in the oceans to 50 km at the continental mountains. Since the crust makes up such a small amount of the planet's mass, it does not figure into calculations of distant Earths.

Next down is the upper mantle, a layer of olivine, pyroxenes and garnet. It is within this region where plate tectonics take place. Below this layer lies the lower mantle, dominated by silicates in various forms. At the planet's center, a solid iron core 2400 km in diameter floats within a molten iron outer core.

Venus is likely similar in structure, as it is Earth's twin in size. Mars affords another example that can be applied to exo-Earths (Earth-like exoplanets). Current models of the Martian interior suggest that its core region is roughly 3590 km across. Data and models imply that the core consists primarily of iron and nickel, with a bit more than 15 % sulfur thrown in. This iron sulfide core is partially fluid, and contains a larger component of lighter elements than Earth's core. A silicate mantle surrounds the Martian core. As on Earth, that mantle is responsible for many of the tectonic and volcanic features on the planet, but it now appears to be dormant. In addition to silicon and oxygen, the most abundant elements in the Martian crust are iron, magnesium, aluminum, calcium and potassium. The average thickness of the planet's crust is about 50 km, thick compared to the overall planet. In contrast, Earth's crust is just one-third as thick—compared to the planet—as the Martian crust.

In the case of Earth, the solid iron core is surrounded by a molten outer region of liquid iron. This liquid metal moves with the turning of Earth and with inner convection, setting up magnetic fields around the planet. These fields, called collectively the magnetosphere, form a protective bubble around Earth, staving off the barrage of radiation pouring from the nearby Sun. The magnetosphere also protects the atmosphere from being stripped away by solar wind (Fig. 2.7).

Venus, Mercury and Mars have far weaker magnetic fields, in varying strengths and locations. The metallic cores of both Mercury and Mars likely cooled early in their histories, freezing into solid rock as the hot young planets settled into their more sedate modern forms. While Venus is a twin to Earth in size, its rotation is very slow, turning one lazy day each 243 Earth days. Since its year lasts for 224 days, its longer daily turn means that the planet

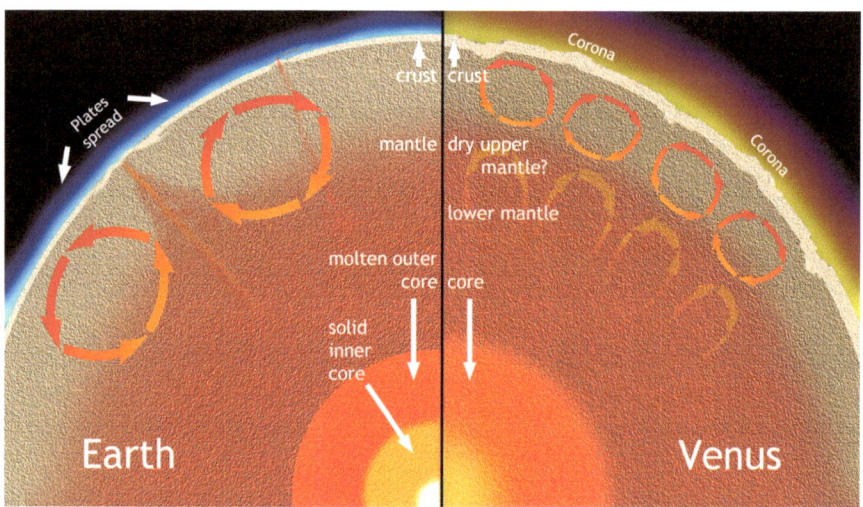

Fig. 2.7 The interior plumbing of Earth (*left*) and its Earthlike sibling, Venus (*right*). (Art © Michael Carroll)

rotates in the opposite direction to most things in the Solar System, in a retrograde motion.[11] The lack of a Venusian magnetosphere may be due, in part, to the planet's slow rotation.

Although a planet's magnetosphere has a dramatic effect on its atmosphere, other factors come into play. In the case of the terrestrial planets, their atmospheres came late. Unlike the gas and ice giants, which drew their atmospheres from the surrounding solar nebula in the Solar System's formative years, the terrestrials lost their early atmospheres as the nearby Sun formed and blew its furious solar winds outward, clearing the inner Solar System of its primordial hydrogen-rich gases. Hydrogen and helium still rule our outer system, but the inner worlds have secondary atmospheres made up of different gases: nitrogen, oxygen and carbon dioxide (in fact, carbon dioxide makes up the majority of gas on both Mars and Venus). As on primordial Earth the new gases of the other terrestrials came from volcanoes, and from the impacts of asteroids and comets. Once in place, the second-chance atmospheres changed and morphed, sculpted by sunlight, chemical interactions with the surface, and loss from solar wind. Earth's third atmosphere, unique in its rich oxygen (O_2) levels, has been altered by biology. As we will see, oxygen may be a "marker" of biology on exo-Earths.

A planet's day/night cycle, seasonal tilt and nearness to the Sun also affect its gas blanket. The slow rotation and dense air of Venus produce high temperatures and pressures. The Venusian proximity to Sol may have triggered a greenhouse effect, vaporizing the extensive oceans it may have had and roasting carbon dioxide from its rocks, adding it to the expanding atmosphere. Mercury and Mars, on the other end of the atmospheric scale, both have low gravity and little in the way of protective magnetic fields, so they have lost more atmosphere over their lifetimes. Their smaller masses meant smaller cores, and as those cores cooled, they shut down the protective magnetic fields. Both planets have had extensive volcanic outgassing, but with little gravity to hang on to what was left, the air around Mercury and Mars drifted away on the solar winds. Venus, Mars and Mercury undoubtedly represent some number of relatively Earthlike exoplanets (Fig. 2.8).

In our search for Earthlike worlds, the past events and climates of our planet inform us that life can exist in extreme conditions, perhaps in exotic locations of other star systems. Nevertheless, the most familiar, Earthlike worlds will be found in a star's habitable zone. Each star type has its own unique habitable zone.

---

[11] Seen from above their north pole, the planets all tend to orbit the Sun in a counterclockwise—or prograde—direction, and they spin in the same counterclockwise motion. Any body moving in the opposite direction has a motion called retrograde.

Fig. 2.8 The most "Earthlike" planets in our Solar System may provide glimpses into what we can expect on some Earths of distant suns. Here, we see three typical surface views of (l to r) Venus, Earth and Mars. (*Left:* Digital revamp of Soviet Venera image courtesy of Don Mitchell and the author. *Center:* Death Valley, California, Earth, photo by the author. *Right:* Curiosity rover image of Gale crater, NASA/JPL/Caltech)

Some HZs are narrow and close in, while others are wide and spread out. The main thrust of our search for Earths of distant suns will focus there. But not all HZs are created equal.

Isn't That Special?

The rich and varied history that our planet has undergone demonstrates a small sampling of the variety we may find among Earthlike planets of other systems. And while much of what we witness in the ancient record here may apply to other earthlike planets, some features may be unique to Earth. For example, the rejuvenating power of plate tectonics, with its recycling of minerals in the rock and gases in the air, does not seem to occur on planets with thicker crust. Our twin in scale, Venus, has no hint of plate tectonics, although it may have a less efficient type of movement within its dense crust. This movement appears to present in a columnar migration of material. Crust moves up and recirculates down in oval regions, as if limited to the outer edge of a pillar.

As evidence of this new type of tectonics, some researchers point to the coronae, vast oval mounds hundreds of km across. Typically, concentric troughs surround the coronae, perhaps formed as upwelling mantle material forces the crust upward. Rather than a surface in constant movement, Venus may endure periodic global resurfacing events. The liquid water on Earth may lubricate the process of plate migration. Without liquid water, Venus is doomed to simmer under a dense atmosphere, warmed not only by greenhouse effects from above but also from trapped heat below, where interior energy is unable to escape. What's left is a world pockmarked by volcanoes, the only clear escape route for its internal heat. The slow spin of Venus also contributes to its alien nature. Without the currents of molten metal sloshing around in its core, Venus lacks a magnetic field. To add insult to injury, its lazy day/night cycle combines with hurricane force winds that blow heated daylight air onto the night side, fostering its global blast-oven conditions.

Unlike Earth, Venus has no moon. Although this might not disturb any Venusian romantic poets, the lack of a moon may actually be an important barrier to life. In fact, our Earth's large natural satellite has a profound effect on the uniqueness of our planet.

Earth's Moon is so large—compared to its primary—that Earth essentially qualifies as a double planet.[12] Luna may have helped to draw away early greenhouse gases, something Venus would have benefited from. It regulates Earth's spin and dampens the wobble of its axial tilt.

Ironically, a violent impact gave birth to our beneficent Moon. Shortly after our planet became differentiated, a cataclysm nearly put an end to the world we inhabit today. The event transpired some 4.2 billion years ago, in the midst of the asteroid demolition derby. Earth's mass became large enough—and radioactive materials abundant enough—to heat the core. But before Earth could settle down into a respectable planet, a Mars-sized behemoth sped out of the darkness. If the angle of impact had been slightly steeper, Earth would have shattered like a dropped wineglass. Instead, the stray planet hit a glancing blow. The colossal collision peeled away the lighter material from Earth's crust. For a brief time in geologic history, the again-molten Earth had a ring to rival even Saturn. Within less than a million years, though, that ring of debris had become our Moon.

The Moon not only keeps our environment stable, it also raises the tides. Many biologists believe the Moon played a critical role in the rise of life. Earthlike worlds lacking a large natural satellite—like Venus—may be at a disadvantage in developing a biome.

One such world is Mars. In many ways, Mars is the most Earthlike world in our Solar System. The planet's tilt is similar to Earth's, providing it with comparable seasonal changes. Its day is only a few minutes longer than Earth's, and daytime temperatures, while chilly, are close to the melting point of water. Mars orbits just at the outer edge of the Sun's habitable zone. But its two moons, Phobos and Deimos, are so small as to have no effect on the

---

[12] While the Moon orbits Earth, the center of rotation for the entire Earth/Moon system is just inside the surface of Earth. The only other two-body system that comes close to this scenario is Pluto, which orbits a point between itself and its large moon Charon.

stability of the Red Planet. This means that over time, the Martian spin axis tips wildly. About 50,000 years from now, the axial tilt of Mars will roll over to the point where the planet spins nearly on its side, creating dramatic—and possibly lethal—seasonal changes. Additionally, Mars' small size means that its core has cooled. Liquid metal in the center has, for the most part, frozen into solid iron, leaving the planet without a substantial magnetic field. Some evidence suggests that the planet once had a robust field, and something like plate tectonics. Because of Earth's tectonics, new rocks are generated on the floor of the Atlantic Ocean, moving away from a central ridge like two conveyor belts sliding away from each other. These rocks retain a record of Earth's magnetic fields, their patterns creating a mirror image of each other. On Mars, hints of this kind of mirror image show up in some areas of the crust, ancient traces of magnetism from long ago. But the core's magnetic field is essentially gone today.

Mars has another problem in the Earth-similarity-department. Its low gravity is unable to hold onto a substantial atmosphere. In similar fashion to Earth and Venus, Martian volcanoes replaced its initial reducing atmosphere with carbon dioxide, nitrogen and water. Because the little world has no shifting plates, some volcanoes became giants, building over a geological heat source for over a billion years. But eventually, the volcanoes all died out, and today so much of the original Martian air has drifted away that pure water cannot exist on the surface; the pressure is so low that it boils away instantly as vapor.[13]

Our Solar System possesses four terrestrial (rocky) planets: Mercury, Venus, Earth and Mars. Of those, only Venus, Earth and Mars could have become habitable. Mercury is not in the habitable zone; it orbits so close to the Sun that daytime temperatures reach 427 °C. On the other hand, both Venus and Mars travel along the inner and outer edge of the zone, affording them the possibility of liquid water under the right conditions. However, they have both lost their magnetospheres, those protective magnetic bubbles surrounding some planets. Without the protection of a magnetosphere, Mars has lost most of its atmosphere, leaving it cold and geologically quiet, too small to have remained active. With its overgrown atmosphere, Venus broils in conditions too hot for life. So although we have three terrestrial planets within a habitable zone, it appears that only one—Earth—sustains a biome.[14]

---

[13] In some of the lowest-lying areas on Mars, liquid water may persist close to the surface, and impurities like ammonia may enable water to retain a liquid state on the surface itself.

[14] Here, we are speaking of life as we know it from the only sample we have: Earth's. Extremophiles living in drastic conditions on our own world demonstrate that life may take many forms in quite hostile environments. The search for life on our nearby worlds, active or in the fossil record, continues.

A Really Big Habitable Zone?

Another limitation to Earthlike worlds may exist: location in the galaxy. It seems that our Milky Way has a habitable zone of its own. The great spiral of stars making up our galaxy spans some 120,000 light-years across, inhabited by 100 to 400 billion stars. Within this structure, interstellar gas and dust drift among the star-birthing nebulae. Young and middle-aged stars glow throughout, but the stars in the central bulge tend to be more geriatric. A supermassive object at the galactic center, known as Sagittarius A*, attracts a maelstrom of gases and stars around it. The object pumps out prodigious amounts of X-rays and may well be a massive black hole.

Our planet orbits the galaxy far enough from the center to have access to the heavy elements coming from exploding stars. Older stars near the center lack many of the heavier elements, while stars out in our neighborhood benefit from heavy elements generated in the explosions of supernovae. This is good fodder for the rich terrestrial planets in our Solar System. At the same time, Earth orbits 28,000 light years from the center, far enough away to elude the lethal gamma radiation coming from the galaxy's heart. Just 5 % of stars in our galaxy fall within this zone.

The Solar System's orbit around the galaxy is also fairly circular, so that it avoids the galaxy's dangerous spiral arms. We have seen that those starry lanes are high-radiation neighborhoods, making them bad territory for life, but there is more to the story. Within those arms, interstellar gas gives birth to new stars whose intense radiation could destroy life on Earth. But our location lets us keep pace with the arms. As a communications satellite orbits far enough from Earth to remain over the same spot, turning at the same rate as the planet, so the Sun orbits at the right distance from galactic central to remain in between spiral arms. Sol's family is also far enough away from the galactic center to miss the disruptive gravitational forces and intense radiation nearer the central hub (Fig. 2.9).

Location of the neighborhood is critical to the existence of Earth-like worlds, but so are the details of each individual planet. What we have learned about our own planetary system can be applied to Earths of distant suns.

Applying Lessons to Exo-Earths

Our studies of members in our own Sun's family make it clear that a planetary interior has a profound effect on the nature of a world. But how can we tell about the interior structure of an Earthlike world we can't even directly see?

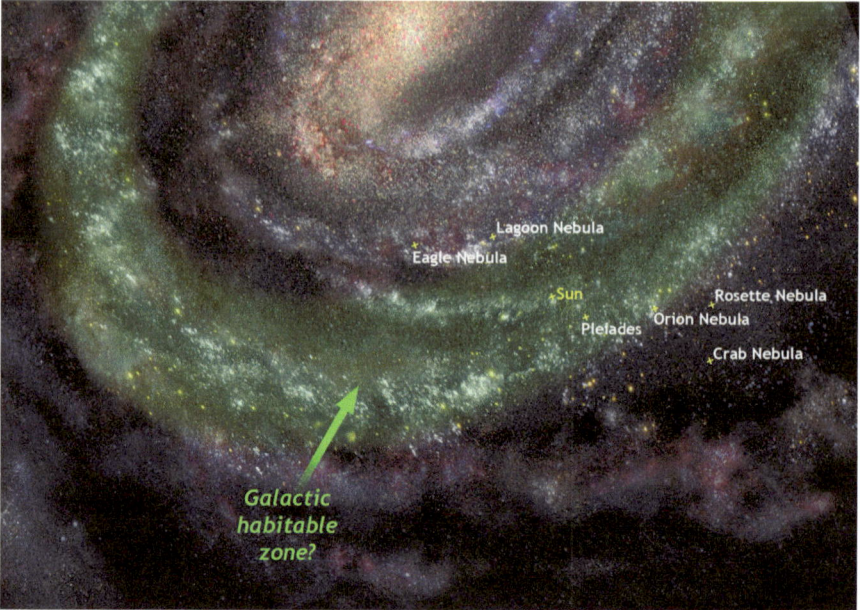

Fig. 2.9 Our place in the Milky Way Galaxy may be particularly well-suited for life, between spiral arms and away from the radiation-filled heart. (Art © Michael Carroll)

A recent study by Caltech's Christophe Sotin et al. indicates that only a few elements are needed to determine the total mass of a planet. It turns out that oxygen, iron, magnesium and silicon together make up 95 % of Earth's bulk. Combined ratios of these elements, compared to the planet's known mass, provide observers with a yardstick for an approximation of masses of distant worlds.

Technology is nearly to the point where we may be able to spectrally tell what these exoplanets are made of, but while we wait for those advances, we can use their estimated masses (found by other techniques such as transits and radial velocity) to approximate the makeup of these faraway worlds. When we do this, we can also make an educated guess as to what the interiors of these exoplanets are like, which tells us about possible magnetospheres, volcanic contributions to their current atmospheres, etc.

Building a model of an exoplanet interior is not as easy as it sounds (and it doesn't sound easy). Adding sulfur, nickel and aluminum to the mix introduces complexities, shifting the melting points and crystalline structure, and changing the model's extent and location of crust, mantle and core. Still, theorists have come up with models that work well with the few terrestrial planets we have as examples. Using only the elements as a guide to the size of Venus, Earth and Mars, the models come within 1 % of predicting how large the planets should be.

We cannot yet determine the specific elements within the crust and interior of exoplanets, but their nearby stars provide clues. Abundances of rock-forming elements such as iron, magnesium and silicon within stars can be calculated by the light spectrum coming from the parent stars themselves. Just how close is that relationship between the makeup of a star and the composition of orbiting planets? The issue is still under debate.

These compositional models supply us with estimates about the three Earthlike worlds of our own system. As we establish how close they come, we can then apply that knowledge to the study of exoplanets. And, as we determine the composition of exoplanets, these models will help us to fill in the blanks, giving us an approximation of what may be going on within these worlds.

3

The Search for and Discovery of Exoplanets

As astronomers became more familiar with the types of stars out there, and they pieced together a sketch of what alien solar systems might look like, it was time to find exoplanets. The hunt was on, but the path ahead was rough and uncertain. How could we find a tiny world floating next to a blazing star from tens or hundreds of light-years away? Was our Solar System the typical model of other planetary families? We needed to study other stars to discover more planetary systems.

At first, telescopes seemed the best tool, and the first place to look was the nearest star system. The nearest star to our own Sun is Proxima Centauri, a red dwarf star in the Alpha Centauri system. Proxima is about half again the size of Jupiter (209,000 km across) and is part of a fascinating triple-star system. It orbits two larger stars that are similar in nature and size to our Sol, Alpha Centauri A and Alpha Centauri B. Proxima orbits a distant 1/5 of a light-year from the other two. It takes Proxima half a billion years to make the long, looping trip once around the system. Proxima is currently on the portion of its orbit nearest Earth, making it the closest star to Earth by a trillion km. To the naked eye, Alpha Centauri A and B look like a single star, the third brightest in the night sky of our southern hemisphere.[1] Proxima is invisible without a telescope.

Alpha Centauri A is slightly larger than our Sun, with a mass about 10 % larger. It is the same stellar type as our own star (G2). Surface temperatures reach 5500 °C (comparable to the Sun's), but its greater diameter, 25 % more than Sol's, makes it 1.6 times as brilliant. Alpha Centauri B is smaller and

---

[1] Only Sirius and Canopus are brighter.

© Springer International Publishing Switzerland 2017
M. Carroll, *Earths of Distant Suns*,
DOI 10.1007/978-3-319-43964-8_3

more orange. Its type is known as spectral type K2 (or orange dwarf). Its lower temperature (5000 °C) results in the star giving off only half the luminosity of Sol. These two primary stars circle each other once every 80 years at an average distance of 11 AU. The red dwarf Proxima Centauri simmers at just 2825 °C, and glows like a dying ember, only 1/500th as bright as our Sun.

Direct telescopic observation can tell us very little about even these, the nearest stars to Earth. The light from a planet's star overpowers the planet's comparatively dim reflected light. An instrument called a coronagraph can block light from the primary star, but this is an exacting technique that is still evolving for exoplanet searches. No planets have yet been 100 % confirmed in the Alpha Centauri system, although a Jovian-sized planet has tentatively been found transiting (moving in front of the face of) the star Alpha Centauri B in Hubble Space Telescope data. But this has yet to be confirmed and is controversial. Planets in such a system could orbit either Alpha Centauri A, B or Proxima, individually, or they might circle in long orbits around both Alpha Centauri A and B, which are as close to each other as Saturn is to the Sun. But it may be that the dynamics of such a complex triple-star arrangement preclude the formation of extensive planetary systems.

There are more stars—and planetary systems—out there. To find them, we must use some very clever techniques.

Finding Invisible Worlds

The search for exoplanets is the search for the hidden. Planets orbiting other suns are lost in the glare of their parent star, concealed in the furious light of day. Directly imaging them is the most difficult way of revealing these distant worlds. Only recently have advances enabled us to find worlds directly. But there are other techniques, and they fall into five categories: radial velocity, astrometry, timing technique, gravitational microlensing and transits.

Direct Imaging

Direct imaging of an exoplanetary system means getting an image of an actual planet in orbit around another star. Planets are far fainter than their host stars and so close to the parent stars that they disappear. Jovian-sized worlds can be imaged more easily than Earthlike globes, but new techniques are in development. In order to directly image distant Earths, the glow from the planet's star must be blocked out. The star's scattered light that would normally blind a telescope's view of nearby planets can be canceled out. One promising technique for this glare removal is interferometry. The SETI Institute's Seth Shostak explains, "If you can link two space-based telescopes in phase, then their view of the sky is a series of fringes, an interference pattern. With correct pointing, you can arrange it so that the star falls in a minimum of the fringe pattern, and pretty much disappears. A planet, however, would *not* be in this [canceled-out region], and its light could be seen. All of this is easier, and more effective, at infrared wavelengths than at visible light wavelengths."

To directly image a distant planet requires a very big telescope. It is inefficient and expensive to build large, single mirrors. But if engineers can craft several smaller telescopes and combine their light—what Shostak referred to as "linking together"—higher resolution can be achieved. In fact, telescopes arrayed together can resolve objects equivalent to those seen by a single telescope covering the entire area *between and including* those arrayed smaller instruments. Some ground-based telescopes today are as powerful as orbiting ones such as the Hubble Space Telescope. However, telescopes that see into the infrared part of the spectrum do better in space, above Earth's atmosphere.

Despite the challenges, directly imaging a planet has distinct advantages over other techniques. A direct image provides researchers with a spectrum, a chart of reflected light, from the planet itself. Spectra can tell scientists what the planet is made of, whether it has clouds, its temperature and how it changes from one observation to another. These important clues help us to paint a picture of a complex world.

Researchers released the first direct images of exoplanets more than a decade after the first exoplanets were discovered. At the time, over 300 exoplanets had been detected indirectly, using radial velocity, transit and gravity lensing methods (see below). By July of 2015, that number had increased to over a thousand confirmed, with another 4696 candidate planets awaiting confirmation.

Radial Velocity

Radial velocity measures a star's movement as its nearby planets tug upon it. Astronomers used this technique to discover the very first known exoplanets. As a large planet orbits its parent star, its gravity causes the star to wobble from side to side. The star also moves directly at—and away from—the viewer. This movement can be measured by a shift in the star's light. As the star is pulled away from us by nearby planets, its light becomes redder, and as its planets pull it toward us, the light becomes bluer. This "Doppler shift" of light can

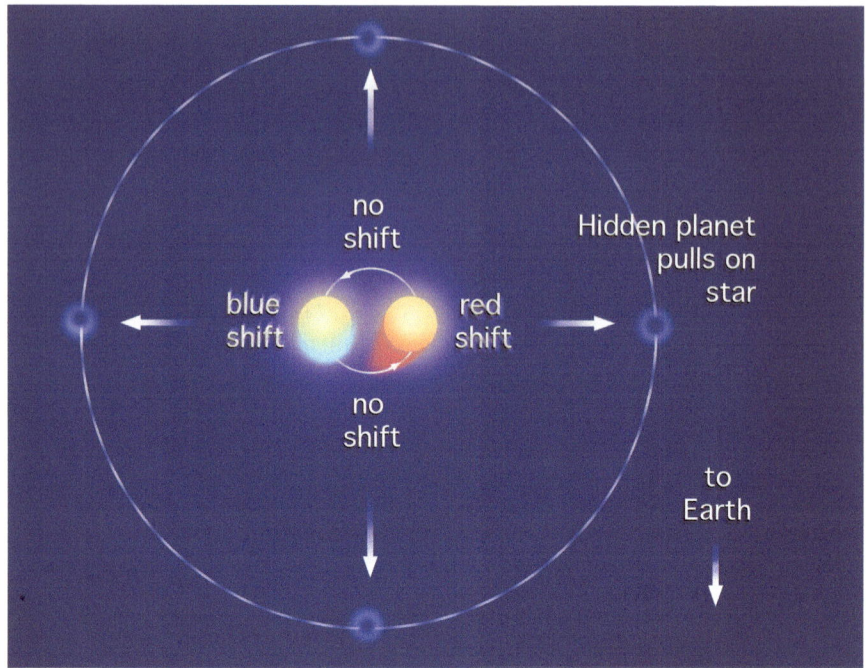

Fig. 3.1 As an unseen planet tugs on its star, the star's light appears to shift toward the blue as it moves toward Earth in its orbit, and red as it moves away. This subtle shift in light reveals the presence of the nearby planet. (Art © Michael Carroll)

be measured in even finer detail than a side-to-side visual movement seen through a telescope. Stars may be far enough away that we cannot see their movements directly, but we can still sense the changes in their light (Fig. 3.1).

By 1953, researchers had totaled up the radial velocities of 15,000 stars, but their instruments were not accurate enough to find the telltale wobbles they were hoping for. As they struggled to refine their equipment, Ukrainian-born astronomer Otto Struve pointed out that the Sun and many other stars rotate more slowly than was predicted by models of stellar evolution. Struve claimed that the reason must be that these stars were surrounded by planetary systems which slowed their spin. The galaxy's great number of slow-spinning stars led Struve to suggest that life in the universe was widespread. He proposed that there might be 50 billion planets in the Milky Way alone.

Radial velocity measurements are well-suited for revealing the mass of planets and their period of orbit (length of year). They have been especially successful in determining the masses of planets at least as bulky as several Earths, and in orbit close to their parent star.

Timing Technique

The timing technique works with stars that vary in regular ways. For example, pulsars are the corpses of collapsed suns. They cast out radio waves as they rotate, like a radiation-beaming lighthouse. These waves are regular and predictable, pulsing out of the star as it spins. Since the rotation of a pulsar is so uniform, researchers can use slight changes in the heartbeat of a pulsar's radio beams to detect any changes in its motion. Just as our Sun wobbles around a point between itself and Jupiter, a pulsar will wobble back and forth as its planets orbit nearby. The shift in timing of the pulses reveals the size of the orbit, its distance from the star, and the approximate mass of the planet. The timing technique is so sensitive that it can measure the effects of a planet down to less than one-tenth the mass of Earth. Researchers can also use the approach to detect multiple planets tugging on each other and the star itself, and unlike some other methods, the timing technique can detect planets that orbit far from their parent star.

Like other techniques, timing a pulsar has its limitations. Pulsars are fairly rare, and astronomers believe that planets around these stars are rare. Pulsars are shrunken cores of formerly gigantic stars, born when a red or blue supergiant explodes in a supernova. What's left is a rapidly spinning neutron star, a collapsed sphere of exotic, neutron-rich material. Its explosion vaporizes any nearby planets. Additionally, in our continuing search for Earthlike worlds, the environment around a pulsar is bathed in deadly radiation, so inhabited Earths are unlikely. Some other types of stars oscillate in predictable ways. In all of these cases, the timing of a star's varying brightness or spin will appear to be slowed or sped up by a planet's passing (Fig. 3.2).

Gravitational Microlensing

Another approach, called gravitational microlensing, takes advantage of a prediction in Einstein's General Theory of Relativity, the concept that mass distorts the space around it. Because of this phenomenon, light bends when it passes near a large mass. This warping of space bends images behind a heavy object such as a star. The effect was first determined by Sir Arthur Eddington and his team of astronomers in 1919, just a few years after the publication of Einstein's theory. By tracking stars as they disappeared and reemerged from behind the Sun during a total eclipse, the observers were able to demonstrate the bending of starlight around the Sun. Observations were made simultaneously from Brazil and from São Tomé and Príncipe, off the west coast of Africa.

Fig. 3.2 A molten moon orbits near a recently destroyed planet, and terrifyingly close to its parent star, a pulsar. Regions around pulsars are unlikely to host Earthlike worlds. (Art © Michael Carroll)

The same phenomenon can be applied to planets circling stars. When one star appears to pass near another (from the viewpoint of Earth), the closer star acts as a lens, increasing the brightness of the farther star. But if a planet is present in orbit around the closer star, a second brightening of the far star will occur. By early 2011, eleven exoplanets had been found using this technique, one of which was, up to that point, the least-massive exoplanet discovered around a main sequence star.[2]

Autocorrelation Function Timescale Technique

Another technique that will aid in the detection of Earthlike worlds is being perfected by a team of French, German and Australian astronomers. The team, led by astrophysicist Thomas Kallinger, is perfecting the "timescale technique," a way to measure gravity's pull at the surface of a star. Armed with this information, researchers will be able to more accurately characterize the planets affecting the star surfaces. Scientists can measure surface gravity of distant suns with an accuracy of about 4 %. A star's gravity depends on its mass and diameter, but many stars are too distant to calculate these. With these remote stars, the dip in light of a transit may be the only hint of a planet, even though the size of the star is unknown. The new technique enables sky watchers to judge the mass and size of distant suns, leading to a more accurate view of the planets transiting them. In the past, several planets were considered Earthlike until it was discovered that their parent stars were larger than thought. This meant that the planets themselves were larger than initial estimates, and thus too heavy to be Earthlike. The new approach will help clarify the sizes of stars and planets in our search for Earthlike worlds. Best of all, the timescale technique uses data already collected from satellites such as Canada's MOST (Microvariability and Oscillations of Stars telescope) and NASA's Kepler Observatory.

Transits

Perhaps the most dramatic advancement in the hunt for exoplanets has come at the hands of the transit technique (Fig. 3.3). By carefully monitoring the light level coming from a star, any drop in that level might betray the passing of a planet in front of it. It's an exacting technique; light levels drop by only 0.01 % to 0.1 %, depending on the size of the planet in comparison to its star. A transit takes place once in each orbit of the planet. The duration of the transit can reveal the planet's distance from the star and orbital speed. The size of the planet can be determined by how much light it blocks out. When

---

[2] Since then, several smaller worlds have been found using other techniques, including Gliese 581c (5 Earth masses) and Gliese 581e (1.9 Earth masses).

combined with the radial velocity method (which determines the planet's mass by how much it tugs on its star), observers can determine the density of the planet, and hence learn something about the planet's physical structure. The planets that have been studied by both methods are the best-understood of all known exoplanets, but they are rare.

A related planet-searching method is called the transit timing variation (TTV) technique. If one planet is discovered by the transit technique, multiple observations of its passing may sometimes reveal variations in the timing of its transits. These variations are sometimes due to other nearby planets that are not lined up to transit the star. This method is so sensitive that planets with Earthlike masses can be charted.

The transit method has two major drawbacks. First, planetary transits only happen with planets whose orbits are perfectly aligned with their parent star from Earth's viewpoint. The probability of a planetary orbital plane crossing its star is the ratio of the diameter of the star to the diameter of the orbit. (In small stars, the radius of the planet is also an important factor.) Only about 10 % of planets in close orbits pass directly in the line-of-sight to their star as seen from Earth. The fraction decreases for planets with larger orbits. For a planet orbiting a Sun-sized star at 1 AU (Earth's orbital distance from Sol), the probability of a random alignment producing a transit is 0.47 %, or about one in 200. This means that stars may have planets present that are still unseen to us, as they do not pass in front of their host star. For example, a planetary system aligned with the star's pole pointing toward Earth will present a solar system looking like a target. None of the planets will cross over the star as viewed from our observatories. Even an extraterrestrial observer looking at our Solar System edge-on might be able to detect only one or two terrestrial planets. Because of the slight inclination of their orbits, alien astronomers wouldn't be able to detect Mercury, Venus, Earth *and* Mars from the same observing point. They are not quite coplanar. But by scanning large areas of the sky containing hundreds of thousands of stars at once, observatories such as the Kepler Space Telescope (see below) can use transit surveys to find extrasolar planets at a rate that beats that of the radial velocity method.

Red giant stars have another issue that throws off planet detection. Although planets around these stars are more likely to transit due to the stars' larger size, their fingerprints are hard to separate from the main star's light curve, because red giants have frequent pulsations in brightness. These can occur within periods of a few hours to a few days. M dwarf stars also "flare" at times, throwing off the light curve. But with enough repeat sightings, astronomers can determine if an observed transit is really from a planet (once in each of the planet's 'year') or simply a false reading from variations in starlight.

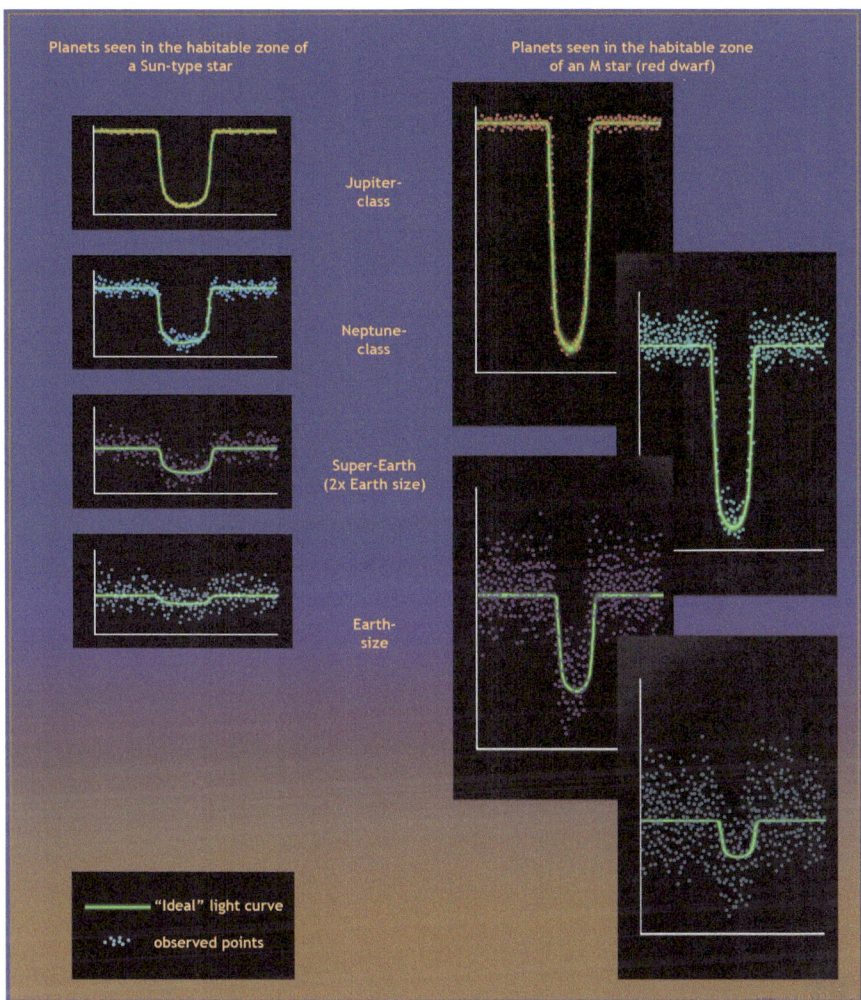

Fig. 3.3 *Left column:* Passing in front of a Sun-sized star, we see the transit light curves from planets the size of Jupiter, Neptune, a super-Earth (twice the size of Earth) and an Earth-sized planet, all in the habitable zone of the star. Had our telescope observed the starlight constantly, we would see the idealized, smooth green line. If we look at the same-sized planets in the habitable zones of an M dwarf star (*right column*), the scattering of the transits becomes far more focused, because the star's face is smaller, so that the planet blocks out more light and is more easily charted. Planets transiting M stars also make more frequent passes, as they orbit closer in, providing us with more observations in a given period. The light levels drop four times as far for the smaller M stars. Finding a true Earth at a sunlike star is much more difficult than finding one in the light curve of an M star, as its passages only occur every year or so. (Author diagram based on work by Thomas Barclay, NASA/Ames)

If an alien civilization gazed at our own Sun using the transit method, would they see us? Their planet would need to be within the same plane as the one in which Earth orbits the Sun, called the ecliptic, so that Earth would appear to cross the face of the Sun with each passage. The distant observers would need to be patient. After spotting us once, alien astronomers would need to wait for 365 days for our stellar light-blip to show up again. But with enough observations, they could tell roughly how large our planet is, its distance from Sol, and its mass. From these elements, they could begin to estimate the nature of our planet, and how closely it resembled their own.

Roughly 80 known stars within 3500 light-years fit the bill for transit observations of Earth. Astronomers Rene Heller (at the Max Planck Institute) and Ralph Pudrits (at Canada's McMaster University) suggest[3] that these stars are the ones we should be paying attention to, as any civilizations within their planetary systems may well know we are here. The stars in the study are similar to—or a bit cooler than—the Sun. They all lie in a region that the authors call the solar transit zone. This is a disk that fans out in the same plane as our ecliptic, which is also the plane of the zodiacal constellations. The study's charted stars are the obvious ones, but many others, including a host of red dwarfs, are probably still hiding. Heller and Pudritz estimate that there may be 300,000 uncharted stars in this region alone, with some 30,000 terrestrial planets in habitable zones. That's a lot of territory, and a lot of possibilities, for advanced civilizations looking out for fledgling civilizations like ours (Fig. 3.4).

Astrometry

Astrometry measures the position and movement of a star against the background sky (in contrast to radial velocity, which uses changes in the color of starlight to sense the movement of a star as it is jostled by nearby planets). Astrometry can fill in the gaps left by other techniques. For example, it can tell observers the inclination of the planet, which helps to determine its mass. The mass provides insights into its bulk composition and surface gravity, which in turn reveals something about the nature of the planet itself: is it a gas giant? A terrestrial planet? Might it have atmosphere? Liquid water?

Although astrometry can chart the movement of our Solar System and other stars in the Milky Way, and determine distances to other stars, the technique

---

[3] Heller and Pudritz, *Astrobiology* June 2015.

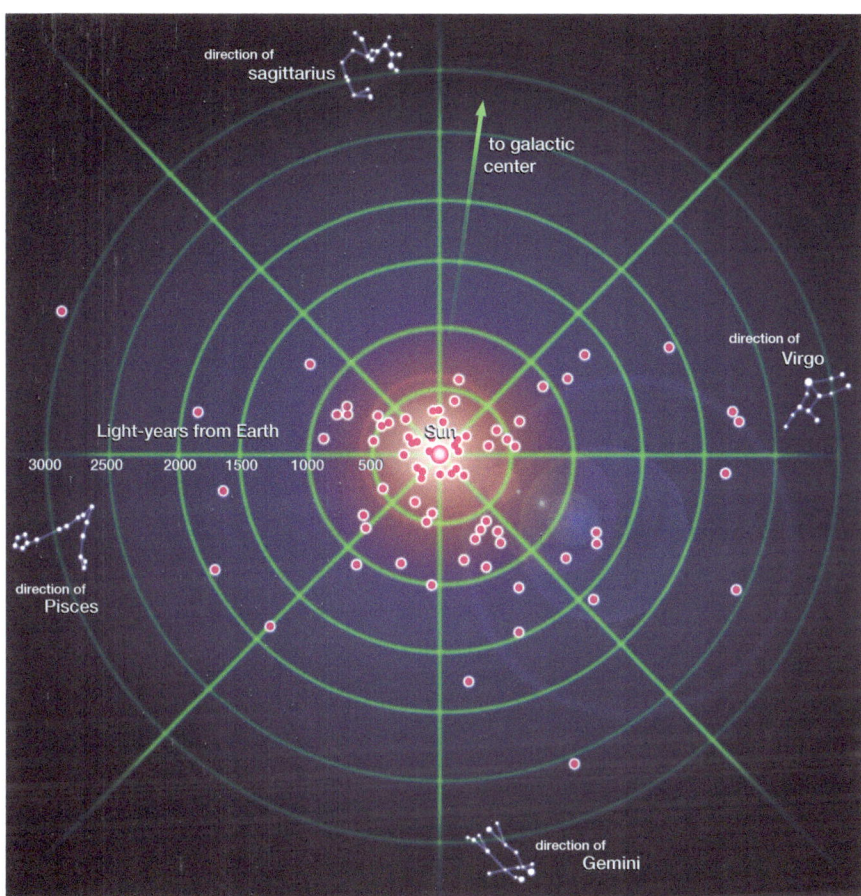

Fig. 3.4 Using the transit technique, astronomers can detect the drop in a star's light by a passing planet. Alien civilizations aligned with our orbital path could see light-level drops from Earth as it passes in front of the Sun. Here, we see 82 known star systems that lie within this thin "transit plane." (Diagram by author, based on work from Otwell/Crockett at *Science News*, and data from R. Heller and R. E. Pudritz/Astrobiology 2016)

has yet to discover exoplanets. But astronomers feel that as they refine the process, astrometry will undoubtedly yield new discoveries soon. In fact, astrometry has confirmed several exoplanets previously discovered by radial velocity. But in order to find Earthlike worlds, instruments will need to be hundreds of times as sensitive as they currently are (Fig. 3.5).

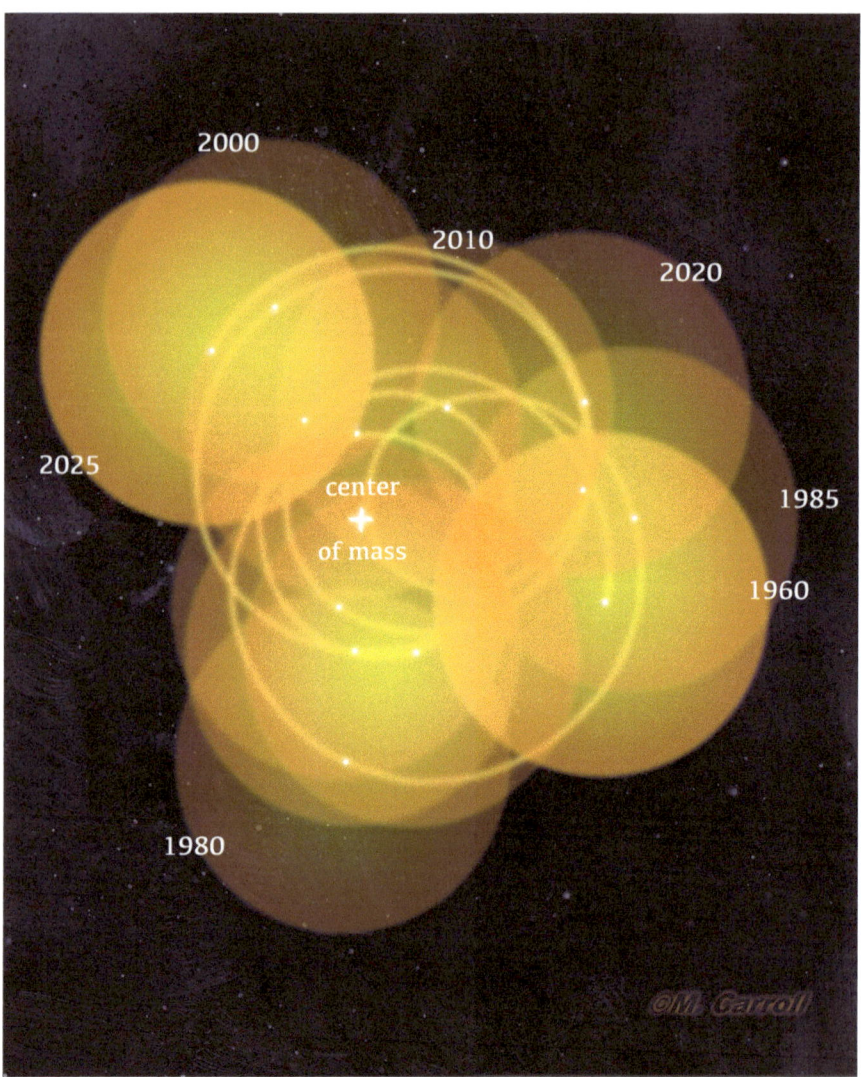

Fig. 3.5 Seen through the telescope of an extraterrestrial observer 30 light-years away, our Sun would appear to revolve around a point between itself and Jupiter. Jupiter's immense bulk causes a visible wobble in Sol's path, charted here at intervals between the years 1960 and 2025. From 30 light-years distance, the subtle back and forth motion of the Sun is equal to one-third the width of a human hair seen from a range of 5 km. (Diagram © Michael Carroll)

The New Age of Discovery: Orbiting Observatories

In 1923, German scientist Hermann Oberth published a paper called "*Die Rakete zu den Planetenraumen*" (The Rocketry into Planetary Science). In it, he advocated the use of orbiting telescopes high above Earth's turbulent atmosphere to increase their resolving power. Later, physicist Lyman Spitzer and Wernher von Braun also sang the praises of the strategy. But the world had to wait until 1990 for a large-scale instrument.[4] The waiting ended with the space shuttle launch of the Hubble Space Telescope. After some false starts and daring orbital repairs by U. S. and European astronaut crews, HST began to return spectacular images far outshining expectations. Oberth was right: an orbiting telescope can see far more than a ground-based one.[5]

Hubble quickly added up accomplishments. It imaged Cepheid variables to determine how quickly the universe is expanding. It studied supernovae and black holes. The spacecraft monitored activity on other planets, and imaged Pluto, Eris, and other remote objects. For the first time in history, Hubble was able to image disks of material around stars where planets were forming.

It was only a matter of time until Hubble joined the search for, and study of, exoplanets. In February of 2004, Hubble carried out a study of 180,000 stars near the center of the Milky Way. The observatory discovered 16 possible transiting planets, all in a class similar to Jupiter. Of those, two have been confirmed, and are the farthest exoplanets yet discovered, at a distance of 21,000 light-years away. The other 14 are too dim to confirm from the ground. If the ratio of 16 planets to 180,000 stars holds, Hubble's scientists estimate that there are 6 billion Jupiters in the Milky Way.

Hubble also combined forces with another discovery. Through the radial velocity technique, ground-based astronomers detected a planet orbiting the star Gliese 876. But their data contained large error bars. The size of the orbiting companion could have ranged from 1.9 Jupiter masses to 100. On the upper end, the companion would have been a small star or brown dwarf, but if it was on the lower end, it would turn out to be a bona fide exoplanet. Hubble was able to resolve the actual movement of the star (astrometry), demonstrating that the orbiting body must be 1.9 Jupiter masses. The success of Hubble has spurred the successful deployment of several other space-borne observatories.

---

[4] HST was not the first orbiting telescope, but for many years it was the biggest and most powerful.

[5] However, some of today's ground-based observatories are getting comparable views using new corrective optics techniques like laser attenuation of mirrors.

The Kepler Space Telescope ushered in a renaissance of exoplanet research. Launched in 2009, the telescope's sole aim was to chart the light levels of 145,000 stars in a small portion of the sky. The craft surveyed the stars in its field of study over and over again, looking for slight variations in starlight. The amount of dip in starlight revealed how much of the star's disk a planet blocked. With careful timing of the planet's orbit by further observation, scientists can determine the size and mass of a planet, along with its orbital characteristics around its star.

Oddly enough, the idea for the Kepler transit mission came from antiquity. The mission and telescope bear the name of sixteenth century astronomer Johannes Kepler, says mission designer Bill Boruki. "Kepler, the astronomer, was broke. Had to support his family. What does he do? He goes to Tycho Brahe, who was a very wealthy person. Brahe had lots of money coming in, he was building his own observatory, and so forth. And Kepler says, 'What do you have for me to do?' Brahe says, 'We're really having a difficult time predicting where Mars transits the meridian. It's supposed to transit at a particular time, and it's off all the time. Why don't you figure out why all the transits are off?' They were doing that back in the fifteen hundreds. It's exactly what we're doing! But it's a great thing to continue because you learn so much. It's a tremendously powerful method."

The Kepler spacecraft carries only one instrument, a device called a photometer used to sense the drop in light levels from stars (the transiting technique). The transiting technique has had limited success on the ground because of the turbulent effects of our atmosphere. But the entire field of research broke open with the launch of the Kepler Observatory. Using this technique, Kepler gazes at the light of hundreds of thousands of stars, unhindered by the effects of Earth's turbulent atmosphere. Although it has been spectacularly successful, the spacecraft would never have existed had it not been for the Herculean efforts of one man, NASA Ames researcher and engineer William Boruki.

"In the eighties," Boruki remembers, "we had a series of interesting seminars. One seminar at the time asked, 'How do you find planets?' It was pretty clear that you do it with astrometry. At the time that was the idea: to watch for the movement of a star. I got very interested in participating, and I talked to one of the people here. They said, 'Why don't you try something else? Read this paper on photometry (the study of light levels).' So I did. The paper was full of mistakes and impracticalities, but the point is it was a logical exposition of how to use photometry to find planets. So it was just a matter of fixing up the science and building a photometer a thousand times better than anyone had ever built." Boruki did just that. He had done similar photometry work on the Apollo project. He put his talents to work on a new generation of

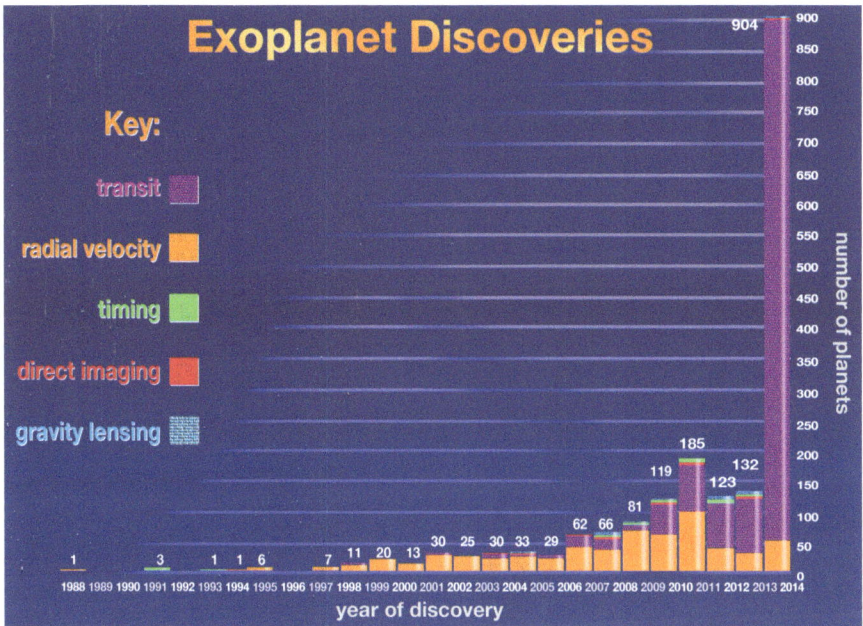

Fig. 3.6   Exoplanet discoveries, plotted by year and technique, increase dramatically as technology improves. Note that the Kepler space telescope, which uses the transit method, came on line in 2009. (Diagram by the author)

instrument, "so we could go and find Earths." The result, after many roadblocks and detours, was a spacecraft that has discovered more planets than all other techniques combined.

The Kepler Observatory has now confirmed roughly 2325 planets in total (Fig. 3.6). The architectures of the solar systems represented by these planets are remarkable in how far they differ from predictions. "One of the most exciting discoveries of the Kepler mission has been that most of the planets we find are nothing like the planets in our own Solar System," Kepler's Thomas Barclay asserts. "There were predictions before Kepler launched that the spacecraft would find no super-Earths (terrestrial worlds weighing at least 1.5 Earth masses). We thought planets about twice the size of Earth just didn't exist. But now we find that the most common planets out there fall into this class."

What would Kepler see if we took Jupiter, the largest planet in our Solar System, and placed it in an orbit around a distant sun-like star? Jupiter's diameter is roughly one-tenth of our hypothetical sun-like star's, which means that its silhouette against the star's disk covers 1/100 of the star's total surface. The level of starlight reaching us would drop by 1 % as Jupiter passes across its disk. A planet the size of Earth is ten times smaller than Jupiter. The starlight from a transit of a dinky Earthlike world would drop by only 0.01 %. Sensitive instruments such as charged coupled devices—the same chips that enable your camera to take photos in dim light—have the capability to chart planets as tiny as Earth when they cross the face of distant stars. Confirmation of the existence of an exoplanet takes multiple observations. If a planet orbits at an Earthlike distance from its star, the probability of seeing it in transit is roughly half a %. If the planet lies closer to the star, the probability increases because of its more frequent transits.

In order to find the subtle variations in starlight, Kepler needed precise aiming and stable viewing. This was done using three reaction wheels, heavy spinning disks that acted as gyroscopes. When one of Kepler's reaction wheels failed, the mission seemed lost. But engineers devised a way of using the pressure of sunlight against the spacecraft's solar panels to stabilize the probe in the axis of the failed wheel. The solution was a complex and elegant longshot. "That's the most fun and the hardest work I've ever done," says Thomas Barclay. "Once [Kepler] failed, some folks wanted to move on because they were very much focused on Kepler's mission goals. But a few of us realized we had to do something with this spacecraft. The folks at Ball Aerospace really pulled this out of the bag. They came up with this idea [of using the pressure of sunlight as a substitute for one of the wheels], and there was a lot of skepticism that it would work, but we were allowed to spend the time and resources to actually test this idea. When it began to work, we said, 'Hey, we can put a mission together.'" Kepler's second chance, called the K2 mission, has resulted in over one hundred further discoveries. Its performance has far surpassed even the most optimistic projections, and the mission continues as of this writing.

In the meantime, ongoing measurements with the ground-based Keck HIRES telescope and HARPS-N (High Accuracy Radial velocity Planet Searcher-Northern hemisphere) will give astronomers the Doppler monitoring they need to determine planets' masses. The European probe Gaia will accurately measure distances to various exoplanets, locking down sizes of planets caught transiting their stars.

In 2017, NASA's launch manifest calls for the liftoff of the Transiting Exoplanet Survey Satellite (TESS). The advanced probe will continue Kepler's work of charting the transits of planets. But unlike Kepler, TESS is tooled to better see transits of smaller planets, more the size of Earth. These planets will orbit nearby stars, close enough to also obtain radial velocities. While Kepler's main mission searched a specific area of space, TESS will survey the entire sky with an eye out specifically for super-Earths and Earthlike worlds. "Kepler has taught us that there are planets everywhere," says Barclay. "The next mission is going to teach us where the best targets are. That's the TESS mission." Another difference between Kepler and TESS comes from its aiming capabilities. Mission planners hope to study the entire sky, repeatedly charting the light curves of 200,000 stars over the course of 3 years. The spacecraft uses four wide-field CCD cameras.

In concert with the TESS operation, the James Webb telescope will study the atmospheres of those targets. Webb carries a coronagraph to block the blinding light of a star, revealing nearby planets for direct imaging. The spacecraft will also carry out spectroscopy of those planets, unveiling the composition of the atmospheres. Webb studies stars and planetary systems in the infrared.

"After that," Barclay says, "we're going to study atmospheres in detail, using missions like the W-FIRST (Wide Field Infrared Survey Telescope) mission in 2024. That's going to find habitable zone terrestrial planets and actually directly image those planets so we can start to model atmospheres. Even beyond that, teams are studying the generation after that, telescopes of the 2030s that will study in depth the properties of these exoplanets."

51 Pegasi b…and Beyond

In 1995, Swiss astronomers Michel Mayor and Didier Queloz found the first confirmed exoplanet, 51 Pegasi b. As are all exoplanets, the name of the star— 51 Pegasi—is followed by the planet, assigned a letter in order of its discovery or distance from its star. The new planet inhabited what was thought to be an unlikely place. It was a Jupiter-sized globe orbiting *51 Pegasi* in a surprisingly close track. Taking only four days to race around its sun, the massive "hot Jupiter's" surface may broil at 1000 °C (1830 °F). The discovery was soon confirmed by Americans Geoff Marcy and Paul Butler. Just 3 years later, investigators using the same technique revealed two more worlds, one orbiting the star 70 Virginis in the constellation Virgo, and another circling 47 Ursae Majoris in the constellation Ursa Major (the Great Bear, which contains the Big Dipper). Since then, radial velocity has proven to be a reliable approach for hunting planets, and especially for finding those extremely close to their stars. Otto Struve would be pleased.

51 Pegasi b gave scientists a taste of things to come, with its close orbit and searing temperatures. Other discoveries soon followed, demonstrating that examples of toxic, deadly worlds abound. Since the discovery of 51 Pegasi b, the detection of exoplanets continued to come in, first at a trickle, and then in a flood. Even before the revolutionary Kepler Space Telescope planet-hunter,

researchers were reporting ten new planets each month, but none of them appeared to be ripe for benign, Earthlike conditions.

The year 1999 saw the first discovery of a multiple-planet system around the star Upsilon Andromedae. Another exoplanet soon followed, this one orbiting 61 Virginis, a star just 28 light-years away. 61 Virginis is nearly a twin to the Sun. Its resident super Earth, 61 Virginis b, weighs in at between 5.1 and 7.5 Earth masses. The planet orbits so close to its sun that it completes each year in just over 4 days. If this behemoth is anything like a gas giant, it is undoubtedly losing atmosphere from the fierce solar wind of its nearby star. A little further out orbits 61 Virginis c, an even larger planet with an orbital period of 38 days. No planets in this system have been found in the habitable zone.

Two years later, researchers finally spotted the first planet within a star's habitable zone—70 Virginis b—and the search for distant Earths took on a more optimistic bent. The existence of a planet in a habitable zone fired the imaginations of the press and public alike. Here could be a world with liquid water, a world that plays host to an environment we might understand on a human level. But the planet weighs in with a mass nearly 2000 times that of Earth, so it is clearly not an Earthlike world. However, astronomers realized that large planets can have large moons, and those orbiting moons might have quite Earthlike conditions. Who knows what kinds of small worlds orbit those gas giants?

With some 6050 candidate planets under their belts,[6] trends are becoming more clear. Several types make up the majority of discovered worlds. Because of the relative ease of finding the largest, the list is biased toward the big planets. Many of the earliest discoveries were the hot Jupiters, gas giants whose mass exceeds half that of Jupiter. Nearly half of the exoplanets with known masses fall into this category. Most have orbits that are dramatically unlike that of Jupiter. These behemoths orbit close to their suns, often more closely than Mercury orbits the Sun. Their typical distances range from 0.015 to 0.5 AU.

Because of their proximity to their suns, these worlds suffer such high temperatures—up to 3600 ° C—that some may grow a comet-like tail as their atmospheres stream off into space, stripped away by the nearby parent star. One such planet is SWEEPS-10 (Fig. 3.7). The giant world careers around its star so closely that it circles once every ten hours. SWEEPS-10 must contain at least 1.6 times the mass of Jupiter, or it would not hold together in its frenzied orbit. Its temperature, estimated at 1650 °C, is hotter than an iron-

---

[6] As of early 2015.

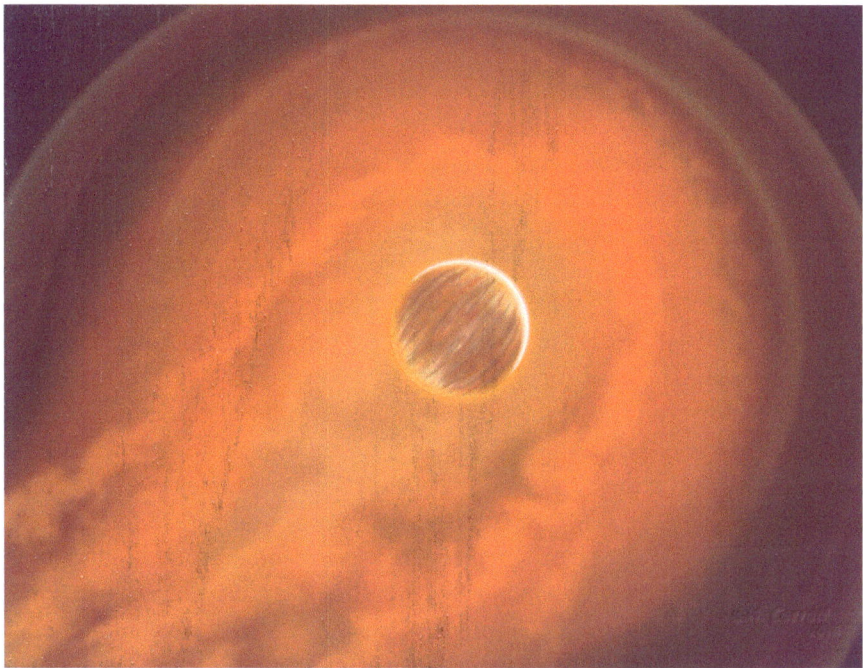

Fig. 3.7 Atmosphere streams from the hot Jupiter SWEEPS-10, stripped away by the nearby star's solar wind. The planet orbits so close to its home star that its atmosphere forms a comet-like tail. (Art © Michael Carroll)

smelting blast furnace. Hot giants range in size and mass from larger than Jupiter to diameters similar to Neptune.

Some of the hot giants would float in a swimming pool. Several hot Jupiters contain less than 20 % the mass of Jupiter. Two-thirds of them weigh in at between a half and twenty Jupiter masses. Astronomers have nicknamed a new class of planet as "Superpuffs." These gas worlds orbit close to their suns, but hold surprising little mass. The fluffy globes probably form farther out, migrating inward, where they lose the cold gases they accreted in the outer, cooler regions of their Solar System. Typically, Superpuffs log a few times the mass of Earth, but they can be ten times as wide, as large as the Sun's gas giants but only 1/100th of their mass.

Slightly smaller than these searing worlds are the hot Neptunes. Like the hot Jupiters, these planets orbit close to their host stars, at less than 1 AU. Unlike the frigid Uranus and Neptune familiar to us, these red-hot giants hold 3 to 20 % of Jupiter's mass.

One recent addition to the exoplanet menagerie lies just 30 light-years away, at the faint red dwarf Gliese 436. A Neptune-sized planet circles this M star, embedded in a gargantuan cloud of hydrogen. Gliese 436's solar wind forces a great tail of hydrogen from the gas planet's atmosphere, creating a comet-like tail some fifty times the size of the star itself. Astronomers using the Hubble Space Telescope spotted the cloud in 2015. It constitutes the largest structure ever seen orbiting another star. The planet has been cataloged as GJ436b. It orbits less than 5 million km from its star—1/30th the distance from Earth to the Sun—racing around it in just 2.6 Earth days. Some researchers theorize that planets like this one, beginning as gas giants near their host star, ultimately become a super-Earth as their atmosphere is stripped away.

The ice giants of our Solar System (Uranus and Neptune) lie in a twilight zone of planets. They are larger than terrestrials, so their makeup is gaseous, like the gas giants Jupiter and Saturn. But unlike Jupiter and Saturn, they are too small to have cores pressurized enough to form liquid-metallic hydrogen at their hearts. Instead, their cores consist of hydrogen, methane, and water, all in the form of dense ices. But the hot Neptunes may not bear much resemblance to the ice giants of our system. Their nature will depend on their background.

"If you have a fairly freshly made hot Neptune," says ice giant expert Heidi Hammel, "like a planet that recently had an interaction with a Jupiter-sized planet that bumped it in close, at first it will still retain all that atmosphere. There may be a period of time where the atmosphere is getting hot and puffing up. It will be like a super-puff before it loses all that material." But the older hot Neptunes may only retain remnants of their dense cores, perhaps transformed into huge balls of rock, metal, and strange conglomerations of other materials. NASA exoplanet expert Mark Marley cautions that hot Neptunes may not appear as Neptune-like as we would think. "In Jupiter's atmosphere, we find these sulfur clouds that are condensed out below the ammonia clouds. But in some of these warmer worlds, the sulfur is in the gas phase. You could have yellow sulfurous clouds, too. Your mental picture is these dark blue Neptunes, but once you heat them up you really can get into weird appearances." Adding to those weird appearances are a host of other possibilities: steamy water vapor atmospheres laced with methane and ammonia; exotic clouds of salt or sodium sulfide; photochemistry in upper atmospheres converting methane fumes into dense smog layers more impressive than those in Beijing; steamy hot.

Chilled Giants

Another frequent discovery has been that of cold Jupiters. These exoplanets orbit at least 2 AU from the star, and are probably quite similar to Jupiter and Saturn in nature.

Even more colossal titans loom out among the stars. The super-Jupiters are gas worlds with at least five times the bulk of Jupiter. Although more massive, super-Jupiters up to a mass of 80 times that of Jupiter are roughly the same size as Jupiter. Their density and gravity increase, but their material continues to compress, so their physical size remains the same. Planets weighing in with more than 80 Jupiter masses have enough bulk to become physically larger. When a planet gains this much mass, it hovers on the border between planet and star. With just a little more mass, interior pressure would trigger fusion in the core.

Most of these planetary types have been found in distant orbits, but some are "hot super-Jupiters," orbiting close in. A recent example is the super-Jupiter Kappa Andromedae b, a goliath orbiting the star Kappa Andromedae. The oversized globe orbits at a distance greater than Neptune orbits the Sun, and its mass is a whopping 13 times that of Jupiter. The planet is so large that it is similar to a brown dwarf, nearly large enough to generate nuclear fusion, and only slightly smaller than an active star (Fig. 3.8).

Earths, Mega-Earths, Sub-Neptunes and Super-Earths

Broadly speaking, the term super-Earth is usually applied to planets that are larger than Earth but still have a rocky surface and a thin atmosphere. The term sub-Neptune refers to a small gas giant. However, there isn't a clear boundary between these two classes of planets, and they are often used interchangeably. The line between the two is blurred even further by the fact that often conditions are not known well enough to determine if a planet is a small sub-Neptune or a large terrestrial super-Earth.

Super-Earths, which include sub-Neptunes, comprise the most common exoplanet type. Roughly three out of ten exoplanet discoveries to date fall into this class. These worlds have no analog in our Solar System. Astronomers class super-Earths by their mass, not by any conditions that the planets themselves might possess. Consequently, these planets may orbit near or far from their host sun, with a wide variety of compositions and natures. The vast majority discovered so far orbit close to their star. Some may be terrestrial and rocky, while others may resemble sub-Neptunes. Super-Earths fall in a mass range with an upper limit of ten times that of Earth. Sources differ on the lower limit, which ranges from about 2–5 Earth masses. Most of these dense worlds are probably terrestrial, similar in makeup to Mercury, Venus, Earth and Mars. Several super-Earths have masses estimated at the upper limit: ten times that of Earth.

Fig. 3.8 Astronomers have now been able to categorize a wide variety of exoplanets. (Chart © M. Carroll)

Astronomers contend that super-Earths fall into four possible categories. Low density planets contain large amounts of hydrogen and helium, and are referred to as dwarf Neptunes, mini-Neptunes or sub-Neptunes. Medium-density Earths probably have water as a major component, making them ocean worlds. Some of these may be completely covered by deep water.

A third type has a denser core than a sub-Neptune, but like a sub-Neptune, it has an extended atmosphere. The extent of that atmosphere will depend upon how distant the planet is from its star. The farther away and the cooler it is, the more atmosphere it can retain (just as Ganymede-sized Titan retains a dense atmosphere while warmer Ganymede does not). Finally, high density super-Earths, known as mega-Earths, probably include major components of rock and/or metal.

Super-Earths are another poster child for planetary migration models—the idea that planetary orbits shift dramatically after the planets form—because it is unlikely that super-Earths can arise in the same place they end up. Heidi Hammel says, "We know that inside Uranus and Neptune there are solid cores. We know that from their gravity fields, precession of the rings, and from the shape of these planets. Let's say we pick up Neptune and move it into the orbit of Mercury. The hydrogen and the helium are going to be blown away. They'll evaporate into space. The next level down are ices, heavier materials than hydrogen and helium. These levels have mixtures of carbon and hydrogen and oxygen and methane. You put that stuff in at Mercury, and that stuff is going to blow away over a very short time. You're left with a super-sized core of iron, nickel, etc. Maybe that's how you get some of these super-Earths 2–3 times the size of Earth."

NASA's exoplanet modeler Mark Marley sees this planetary transformation in more culinary terms. "These planets are like onions with layers of clouds. We only see the outermost layers. But as you bring them closer to their sun, the layers of the onion peel off as it gets hotter and the cloud evaporates. Neptune has these methane clouds. If you move Neptune in closer, the methane evaporates. At a Saturn distance you get ammonia clouds. Closer still, the ammonia evaporates, and you have water clouds. Then closer in, the water clouds evaporate." This series of transitions could, under the right circumstances, leave a Neptune-like world as a giant terrestrial planet.

Smaller on the scale of Earthlike worlds are the exo-Earths, rocky planets similar in size and mass to Earth. Their orbits also vary widely, creating surface conditions ranging from hellish to possibly life-sustaining. The most recent data from the Kepler Observatory offers astronomers an updated estimate: there may be as many as 40 billion Earth-sized planets orbiting stars within habitable zones. Many orbit red dwarfs (the most common) or other stars different from our own, but in our galaxy, up to 11 billion Earths circle stars very much like our Sun.

Even among worlds close to the size and mass of Earth, the "Earthlike" pickings appear to be slim. Most orbit too close or too far to retain liquid water and terrestrial-like atmospheres. Typical of these is the nasty Earth-sized planet circling the star Gliese 1132, some 39 light-years from here. Based on the planet's transits and the radius of the star, researchers calculate that GJ 1132b is 1.2 times the size of Earth. From radial velocity measurements of the parent star, they estimate the planet's mass to be about 1.6 times that of Earth. The planet is on the border of being rocky or sub-Neptunian. Its estimated surface temperature broils at that of a self-cleaning oven, about 225 ° C.

The planetary systems around other stars differ markedly in their arrangements, especially when compared to the planets around our Sun. These seemingly unusual planetary arrangements tell us much about the history of planetary formation, and what we might expect in our search for Earths of distant suns. We explore these exotic groupings in the next chapter.

4

Strange Solar System Architectures

Based on the size and arrangement of planets in our own Solar System, we might expect to find far more numbers of humble, terrestrial-sized worlds huddling around other stars, with huge gas or ice giants orbiting farther out. But our telescopes and spacecraft have revealed a surprising plethora of orbital architectures beyond our own familiar planetary system.

In 1952, Otto Struve published a paper declaring that Jupiter-sized planets might orbit as close to their parent star as 0.02 AU. In other words, Struve's estimate put possible gas giants at less than 3 million km from their parent star. This was a remarkable claim, because at the time most cosmologists believed giant planets all formed at about Jupiter's distance from our own Sun, 5.2 AU away. As it turns out, Struve was right. In 1995, some 40 years after his prediction, advanced radial velocity methods unveiled the first exoplanet at a star similar to our own. Jupiter-sized planet 51 Pegasi b circles its sun more closely than Mercury orbits ours, hinting that other planetary arrangements might be quite different from our own.

The Kepler telescope's discoveries added force to the argument that our Solar System did not have a typical arrangement, Bill Boruki says. "The biggest surprise of all was to see this huge variety of planets and planetary systems. Everybody knows how our Solar System was formed. It's hot near the star, so the only thing that can come out are little rocky planets. Far out, in big orbits, you can clomp on to lots of stuff. It's cold so you get Jupiters. But that's not what we found. And we thought the orbits would have to be circular or they'd be running into all kinds of things. But they're not. And define Jupiter. Would you dare have Jupiters in a four-day orbit? Impossible! So all we've seen is surprise after surprise after surprise."

© Springer International Publishing Switzerland 2017
M. Carroll, *Earths of Distant Suns*,
DOI 10.1007/978-3-319-43964-8_4

Caltech astronomer Mike Brown agrees. "The majority of stars that we've looked at have a sort of Earth or twice-mass-Earth planets close to their stars. Some have hot Jupiters, big rocky or small gassy planets really close in. What we've never really found is something like the Solar System. We thought every other planetary system would bear some resemblance to ours. We thought there would be a big Jupiter-sized planet that formed right about where ice can condense, and that's why Jupiter is so big, everything would be on these circular orbits, you would have rocky planets on the inside, gassy planets on the outside. It all made such perfect sense. Once we knew more we realized we can't make sense of it at all. I'm still astounded; all these other planetary systems seem so weird."

As it turns out, some planet-hunters had predicted this assortment of planetary arrangements. Dynamicists realized that the planets in our own Solar System may not have started out where they ended up. Planets, it seems, are slippery things, forming in one place and shuffling off to another. The arrangement of gas and ice giants outside of the smaller terrestrial planets may not have been the norm throughout our Solar System's history. The idea of planetary relocation fits into several theories, including such ideas as the Nice Model and the Grand Tack. These revolutionary concepts came from an unlikely place: the study of the Kuiper/Edgeworth Belt.

The Kuiper Belt begins where the planets end. It stretches from the orbit of Neptune (at 30 AU) all the way out to 50 AU. Beyond the concentric orbiting planets and drifting asteroids of our solar family, it forms a great donut of frozen planetoids and comets. Kuiper Belt Objects (KBOs) orbit in elongated ellipses rather than the fairly circular orbits typical of the planets. The KBO population contains at least 70,000 worldlets with diameters larger than 100 km. We get a hint of the slippery history of our planetary system from the Kuiper Belt itself, where many objects follow paths that are "in resonance" with Neptune. This means that for each three times that Neptune circles the Sun, a more distant object "in resonance" will orbit it exactly twice. Pluto is one of the objects in resonance with Neptune. Computer models reveal that the only way to trap Kuiper Belt objects in resonances with Neptune would be if Neptune started out closer to the Sun than it is today, and then slowly moved outward. Acting like a cosmic snowplow, Neptune would have spiraled out from a close orbit of Sol, shoving a wave of KBO's ahead of it. The Kuiper Belt observations gave rise to an important new idea: the concept of planetary migrations.

A theory called the Nice model (named after the city in France) describes the planets' orbits as morphing from what they once were. The sizes of the orbits are changing. Simulations indicate that sometimes the planets actually

Fig. 4.1 The Grand Tack model of Solar System formation. Jupiter and Saturn migrate inward, and then out again, scattering and mixing the rocky (*brown*) and icy (*blue*) asteroids. *Top row:* Jupiter and Saturn form within 6 AU of the Sun, with rocky asteroids in the inner Solar System and icy ones farther out. *Row 2:* Jupiter migrates inward—with Saturn following—redistributing the asteroids. *Row 3:* Saturn's gravity pulls Jupiter back toward outer system, while the orbits of Uranus and Neptune drift outward; icy bodies are cast inward. *Bottom Row:* Jupiter and Saturn settle into their modern locations, having cleared out most asteroids and sent the rest sunward toward the terrestrial planets. Note that in the last column, Uranus and Neptune are off the page at the far right. Planets and distances are not to scale. (Art © Michael Carroll, adapted from exoclimes.com/Professor Frédéric Pont)

pass into a resonance with each other, so that Jupiter and Saturn might get into a mean motion resonance where the ratio of the orbits is 2:1 or 5:2. If that happened, the Nice model argues that the shifting arrangement would shake up the order of the Solar System, giving rise to the Solar System arrangement that observers see today (Fig. 4.1).

Another model brings more detail to the picture of shifting worlds. Called the Grand Tack, the model projects that, in the past, Jupiter and Saturn marched in toward the Sun, at which time Saturn pulled Jupiter back from sudden death at the Sun. According to the Grand Tack, Jupiter robbed Mars and its surroundings of icy, planet-building material—asteroids that formed outside the habitable zone—sending it toward the inner system.

Thanks to Jupiter's migration in the early Solar System, about one out of every hundred C-type (water-bearing) asteroids were scattered about to the outer fringes of the Asteroid Belt. But for each one of those, at least ten spiraled sunward, delivering water to the terrestrials, where they ultimately became Earth's oceans. The Grand Tack version of planetary history has the advantage of explaining the diminutive size of Mars, the structure of the Asteroid Belt and the birth of terrestrial seas at the hands of incoming space rocks. It also may explain why astronomers are discovering so many giant exoplanets so near to their parent stars.

Astronomer Heidi Hammel says, "You always start simple. Thirty-five years ago we had a fabulous story of how our planetary system formed, before we knew about planetary migration. The story was that the rocky ones formed close in, and the gaseous ones were farther out. That was a fabulous explanation, because obviously that's what we had. Then, when we started finding Jupiters in at the orbit of Mercury, and guess what? That wonderful story of how planetary systems formed had to be completely rewritten. There are going to be many different pathways to exoplanetary systems, to water worlds, to hot Jupiters."

The concept of planetary migration changed astronomers' views about planetary composition. Because planets do not form in place, their chemistries have changed over time. Most migration models describe Neptune-class members of a solar system as moving outward, where they plow into their solar system's Kuiper/Edgeworth Belt. This remote region contains billions of comets rich in volatiles such as water and methane. During the process, planets pick up icy material, adding it to their bulk. In our own Solar System, we may see the result of such a process. Neptune itself may contain an atmosphere today that is enriched in oxygen beyond what it should have from its surroundings. "It's still controversial," says Hammel, "but this may indicate that there's a source of oxygen that's come from outside of Neptune itself. By implication it's been emplaced there." The seeding of outer planets with cometary material got a boost in 1994, when Comet Shoemaker/Levy 9 slammed into Jupiter, causing small but detectable changes in the planet's upper atmosphere. The event led credence to idea that materials can be emplaced by outside sources as a planet changes orbit (Fig. 4.2).

Migrating Planets and the Search for Life

In a star system such as that surrounding an M star, or red dwarf, big planets often circle close to the star. But when a planetary system forms, most of the nearby available water boils away, forced toward the outer reaches of the system. At a certain distance from the Sun's heat, water is able to condense, turning into ice. In this region, water and ice are incorporated into planets, where they can become giants similar to the gas and ice giants of our own

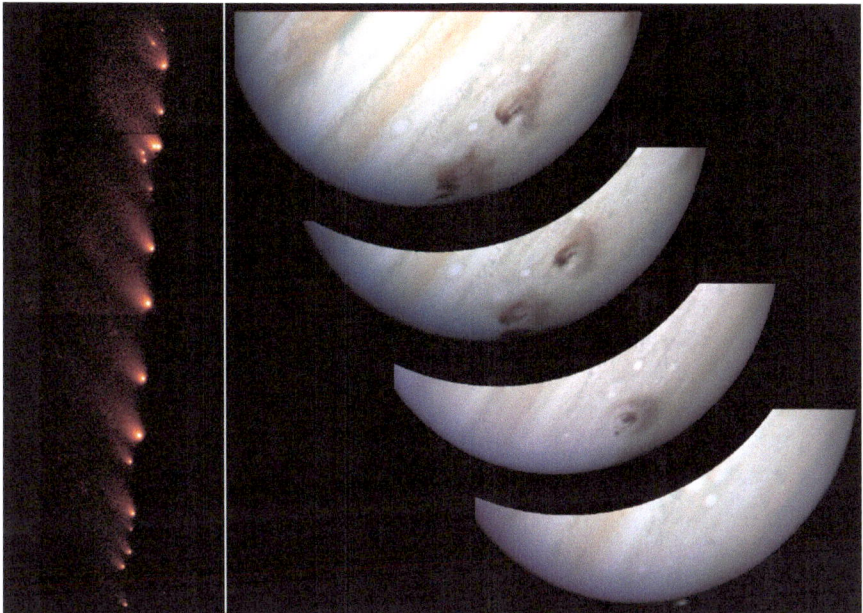

Fig. 4.2 The nucleus of Comet Shoemaker/Levy 9 split into at least 22 separate fragments (*left*, in false color) before crashing into the atmosphere of Jupiter. Hubble images showed Earth-sized blemishes after the explosions. Volatiles from the comet contributed to the chemistry of Jupiter's upper atmosphere; this probably happens in other planetary systems as well. (HST images courtesy R. Evans, J. Trauger, H. Hammel, the HST Comet Science Team and NASA/ESA)

system. The line at which this happens is called the snow line. Inside of it, in the inner planetary system, water becomes more and more scarce. In effect, an inner solar system is a planetary desert, while water in regions farther out becomes more abundant. Entire globes of frozen water circle around the gas and ice giants, and oceans far more vast than Earth's exist beneath the surfaces of some of the moons in the system. But what about all those behemoth planets orbiting close in to their stars? They are clearly inside of the snow line now, but their formation had to take place farther out. This means that even worlds within a planetary "desert" region can retain vast inventories of water, making life more likely.

Adding to the Grand Tack model of planetary migration, new evidence has been uncovered of yet another giant world in our Solar System, one outside the realm of the Kuiper Belt. Because of the clustered orbits of several KBOs, it now appears that a planet with ten times the mass of Earth—nearly the size of Neptune—orbits the Sun at great distance. The chilled giant is seven times as far from Sol as Neptune (200 AU) at its closest, but its oblong orbit may carry it out as far as 600 to 1200 AU. The existence of such a distant world in our planetary family is one more characteristic of migrating planets. Close-in giant worlds such as 51 Pegasi b are yet another. With the knowledge of planetary migration, we can expect to find planets in orbits once thought too exotic to occur.

In commenting about the huge planets found orbiting near stars, planet-hunting astronomer David Jewitt says, "They do indicate the radial migration of planets in other systems. Nobody thinks you can form a giant planet that close to a star, so the idea is that they form far away and they've drifted in… migration of planets is important."

Among the many planets orbiting in their inner solar systems, some orbit too close for comfort and might eventually be torn apart by their star. One such unlucky world showed itself in the subtle ebb and flow of light reaching the Kepler Observatory. As Kepler observed a small white dwarf star, astronomers noted a drop in light that was different from a normal transiting planet. Rather than a bowl-shaped light curve as the planet alternately blocks out more and then less light as it passes in front of the star, this light curve was not symmetrical. The reason—the little planet was dragging a tail of debris as it disintegrated. The tiny planet, only as far across as the state of Texas, marks the slow death of a solar system around a burned-out sun. Its discovery also solves a mystery about white dwarf stars. Because of the path of a white dwarf's stellar evolution, its surface should be devoid of heavy elements. But the light from some white dwarfs betray the presence of elements such as magnesium and iron in their atmospheres, elements that should be long gone. Now, researchers believe these elements are traces of dying worlds, their crumbling globes raining down onto the star's surface.

The original size of the dissolving planet is unknown. Caltech astronomer Mike Brown comments, "The most common types of planets being discovered are these things between 2 and 5 Earth masses. We don't even have anything like that in our Solar System. We don't know what that is. We don't have the most common type of planet, and we don't have the most common locations of planets."

Hammel adds, "The neat classification system that we have in our Solar System—of rocky terrestrial planets with virtually no atmosphere, medium-sized Uranus and Neptune-style planets with moderate atmospheres, and the gas giants like Jupiter and Saturn, the sort of three classification scheme—that's not what we're seeing in exoplanet systems. Instead, we see a whole continuum of different sizes and planetary characteristics."

The plethora of oversized Earths intrigues Kepler Observatory's designer Bill Boruki. "A really great, fundamental surprise has been the planets between the size of Earth and the size of Neptune. These may represent a kind of planet that's never been seen before—an ocean planet. That's a nice addition to our understanding of planetary structures."

In the data to date, Neptune and Jupiter class worlds seem to be at least as common to inner solar systems as terrestrial planets. But while exoplanet hunters have only found a handful of worlds similar to our own, they may eventually find Earth twins in another place—as moons orbiting giant planets in habitable zones.

With all the discoveries of hot and cold Jupiters and Neptunes orbiting close to—or at great distance from—their primary star, other planetary systems seemed to be quite unlike our own. It was not until 2015 that researchers discovered a solar system that appeared to have a layout similar to ours. HIP 11915 is a G-type star similar to Sol. It lies 186 light-years away, and has a gas giant located 4.8 AU from the star. In comparison, Jupiter orbits the Sun at a similar distance of 5.2 AU. HIP 11915 b has a mass that is 93 % that of Jupiter. As the planet resides in its star's outer planetary system, it may share a history of planetary migration similar to that endured by Jupiter. And since models of Jupiter's movement (like the Grand Tack or Nice models) indicate that it paved the way for the formation of terrestrial planets here, a similar situation may exist at HIP 11915. HIP 11915 may turn out to be the first true solar system analog yet found. Its status as a rare example underscores the uniqueness of our own planetary system's architecture among the thousands of stellar families charted so far.

5

Zeroing in on Earth 2.0

Because of the limits in our search techniques, the data tends to be skewed toward larger planets, worlds very different from our own. Many range in size from Neptune-like to Jupiters on steroids. Neptunian and sub-Neptunian planets total greater numbers than any other size of exoplanet discovered so far. Our search for Earth-like worlds might seem frustrated by all of these migrating behemoths, but such is not the case. Migrating megaworlds seem to pave the way for smaller planets in inner systems where habitable zones lie.

Sub-Neptunes

The ubiquitous mini- or sub-Neptunes join the exoplanet list at a mass range up to ten times that of Earth. They are of similar mass to our own Uranus and Neptune. In comparison, Uranus contains 14.5 Earth masses, while Neptune comprises 17. Super-Earths larger than 1.6 Earth radii appear to contain substantial amounts of hydrogen and helium, transitioning from terrestrial to gaseous.

Sub-Neptunes may represent a wide range of planetary types. Research scientist Mark Marley spends much of his time pondering sub-Neptunes. He models the atmospheres of exoplanets at NASA's Ames Research Center. Marley believes that planets of the sub-Neptune class may turn out to be the most varied of any size worlds. "By the time you get to Jupiters, they may turn out to be fairly conformist and uniform. You get bigger than a Saturn or so, and they all tend to be about the same size because they are dominated by their hydrogen/helium atmospheres. Then, when you get down closer to

© Springer International Publishing Switzerland 2017   **75**
M. Carroll, *Earths of Distant Suns*,
DOI 10.1007/978-3-319-43964-8_5

one-Earth mass, they're probably all rocky worlds with a little bit of atmosphere. But this range of Neptune-plus-or-minus—in between—there's probably a huge range of what these planets could be like. It's a range where every single one is going to be unique."

Although sub-Neptunes measure in at a transitional size between small rocky terrestrials and gaseous Neptune or Jupiter classes, their natures may vary wildly, depending on a host of factors. "It's helpful to think about how they got there," Marley explains (Fig. 5.1). "If you scale up Earth, it can hold on to more atmosphere. Above the mass of a few Earths, the size of these things tends to pop up because they're starting to become big enough to hold on to big hydrogen and helium atmospheres. If you turn up the mass knob, you start to get thick atmospheres and the surfaces would slowly disappear under the atmosphere, and you would transition into this Neptune-like world. There's a rocky core covered by an ice and water layer, but the atmosphere is so thick that the surface is at thousands of bars pressure. You'd never see it, and it wouldn't be habitable. Somewhere in the galaxy, is there a 3-Earth mass Earth that really is just a big Earth? Is there a 2-Earth mass Earth that's just a sub-Neptune? The lines are probably fuzzy, depending on the individual history of each world, what it's like in this in-between range."

NASA/Ames' Elisa Quintana has been working with a team attempting to figure out when a planet transitions from being Earthlike to being a gaseous sub-Neptune. "Before we knew of any exoplanets, we had a basic mass/radius relationship based on our Solar System. Kepler only gets size, not mass, so we

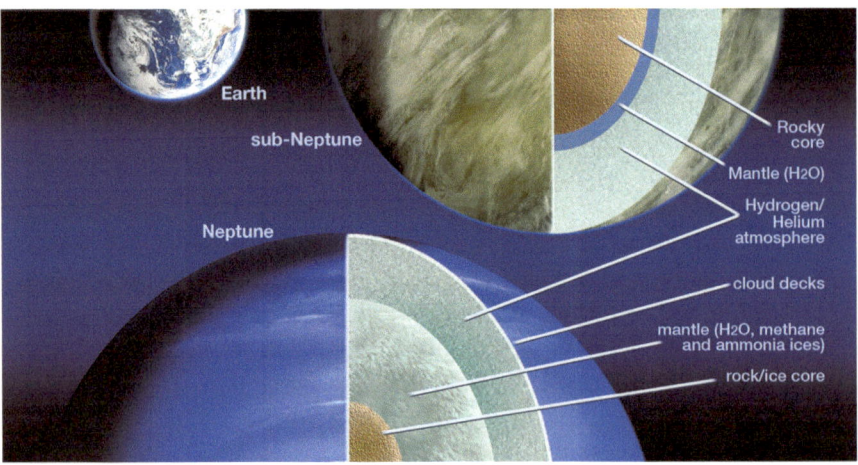

Fig. 5.1 Earth, Neptune, and a sub-Neptune compared. (Diagram © Michael Carroll)

would use this rough ratio. Now, we've had to throw it away. In reality, out of all the confirmed planets, we only have a few with both masses and radii. Data points in that regime are scarce, but what people have figured out using theoretical models is that the transition from rocky super Earth to gaseous sub-Neptune is about 1.5 or 1.6 Earth radii. Once a planet reaches 2 Earth radii, it will be more like a sub-Neptune." Researchers hope to more closely determine the transition point as they study more super-Earths. "We won't really nail this down until we observe more planets with K2 and TESS that are small and can be followed up with radial velocity," Quintana says.

In preparation for these future explorers, astronomer Leslie Rogers of the California Institute of Technology has been studying[1] the elusive transition using Kepler data and radial velocity measurements from ground facilities. Astronomers use the transit technique to determine a planet's radius, in concert with radial velocity or transit timing techniques to understand its density. Armed with the combined data, astronomers have in hand a critical measure of a planet's nature. If, for example, a planet weighs twice as much as Earth, but is the same size, it must be very dense, and so it is rocky. But if a planet of the same mass is ten times the size of Earth, it must be a low density, fluffy world more like a gas or ice giant. Rogers and others are turning their gaze to sub-Neptunes, larger than Earth and with varying densities. These hybrid worlds hover somewhere between rocky exoplanets comprised mostly of iron and silicates, and worlds with extensive atmospheres made up mostly of hydrogen and helium with underlying ices or water. The masses of over 200 transiting planets have now been recorded. Rogers' models indicate that planets with masses over 1.6 that of Earth will rarely be rocky, but instead take on the mantle of a sub-Neptune. One such planet is Kepler 10c. The planet has 15–19 times the mass of Earth, and a substantial diameter of twice as far across as our own world. Its size means that all its mass is spread out into a globe so large that it must be a Neptune-like planet. Kepler 10c orbits a sun-like G star every 45 days, qualifying it as a hot sub-Neptune.

Every rule has exceptions. The super-Earth BD + 20594b has 16 times the mass of Earth, and roughly the diameter of Neptune. Its numbers come in right on the edge of all the models. Most researchers contend that BD + 20594b's mass and radius demand a terrestrial nature, making it the largest solid-surfaced exoplanet yet found. Its close orbit results in surface temperatures of approximately 128 °C. A smaller cousin, super-Earth K2-3d, has a density greater than iron and a diameter half again that of Earth. Although

---

[1] For more on this, see "Most 1.6 Earth-Radius Planets Are Not Rocky," Leslie A. Rogers; *Astrophysical Journal*, March 1, 2015.

its average density is that of the rusty metal, its surface may contain lower density silicates. Its orbit lies in the vicinity of the HZ for its star—possibly on the inner edge—so the planet may have liquid water on or under the surface. But with hotter temperatures, its surface conditions might range more toward the Venusian, with only water vapor locked in a dense atmosphere. Its true nature is still unknown.

Like Neptune, the vast majority of sub-Neptunes are gaseous, with dense, hydrogen/helium-rich atmospheres. Unlike Neptune, these super-Earths tend to stay near to their primary star. Being warm versions of gaseous worlds with no solid surface, we have no analog in our Solar System from which to draw a comparison. The gaseous nature of sub-Neptunes causes exoplanet experts to wonder: how long can a dwarf Neptune last so close to a star? Not for long, as it turns out. Sub-Neptunes must form in the outer regions of their planetary systems, beyond the snow line, where cool temperatures enable them to collect ices and gases—primarily helium and hydrogen—in large quantities. Consequently, they must migrate in later. In a process called hydrodynamic escape, a sub-Neptune's star heats its bulky atmosphere. Lighter gases such as hydrogen and helium expand, and the molecules of gas speed up to supersonic velocities. The movement is enough to accelerate the gases to escape velocity, and the solar wind carries it off.

What's left of the sub-Neptune is the core region, a sphere of water and rock that can take on a more familiar terrestrial form. Astronomers refer to a leftover planet like this as a habitable evaporated core (HEC). Because of their genesis beyond the snow line, many HECs will have enough water to inundate any continental areas. The result is an ocean world with abyssal seas—a globe completely devoid of dry land. An Earthlike world that forms so close to its sun will have a much higher density, because it will lack the water content of an HEC. As researchers begin to track down the densities of exoplanets, the numbers will help them discern whether an Earth-sized world formed in place or migrated there from the outer system.

What's in a Name? The Monikers of Exoplanets

The Executive Committee Working Group of the International Astronomical Union recently asked the public to submit names for a list of 31 new exoplanets and their 14 stars. Until now, planets were named after their primary stars and assigned a letter. For example, the hot Jupiter planet orbiting 51 Pegasi was called 51 Pegasi b. It is now known as Dimidium (Latin for "half," because its mass is estimated to be at least half that of Jupiter's), and its star has been christened Helvetios. Star names in (parentheses) have common names from antiquity.

Star name	Original star designation	Planet name	Original designation
Veritate1	4 Andromedae	Spe	14 Andromedae b
Musica	18 Delphini	Arion	18 Delphini b
Fafnir	42 Draconis	Orbitar	42 Draconis b
Chalawan	47 Ursae majoris	Taphao Thong	47 Ursae majoris b
		Taphao Kaew	47 Ursae majoris c
Helvetios	51 Pegasi	Dimidium	51 Pegasi b
Copernicus	55 Cancri	Galileo	55 Cancri b
		Brahe	55 Cancri c
		Lippershey	55 Cancri d
		Janssen	55 Cancri e
		Harriot	55 Cancri f
(Ain)	Epsilon Tauri	Amateru	Epsilon Tauri b
Ran	Epsilon Eridani	AEgir	Epsilon Eridani b
(Errai)	Gamma Cephei	Tadmor	Gamma Cephei b
(Fomalhaut)	Alpha Piscis Austrinus	Dagon	Alpha Piscis Austrinus b
Tonatiuh	HD 104985	Mextli	HD 104985 b
Ogma	HD 149026	Smertrios	HD 149026 b
Intercrus	HD 81688	Arkas	HD 81688 b
Cervantes	Mu Arae	Quixote	Mu Arae b
		Dulcinea	Mu Arae c
		Rocinante	Mu Arae d
		Sancho	Mu Arae e
(Pollux)	Beta Geminorum	Thestias	Beta Geminorum b
Lich	PSR 1257+12	Draugr	PSR 1257+12 b
		Poltergeist	PSR 1257+12 c
		Phobetor	PSR 1257+12 d
Titawin	Upsilon Andromedae	Saffar	Upsilon Andromedae b
		Samh	Upsilon Andromedae c
		Majriti	Upsilon Andromedae d
Libertas	Xi Aquilae	Fortitudo	Xi Aquilae b
(Edasich)	Iota Draconis	Hypatia	Iota Draconis b

Sea World: A New Kind of Planet

One of the greatest Hollywood disasters in recent history was the post-apocalyptic 1995 tale *Waterworld*. Set in a future where Earth's polar caps have melted, inundating the world's dry lands, Kevin Costner's big-budget flop portrayed a planetary landscape that may be surprisingly prevalent among exoplanets.

The "ocean planets" that so intrigue Bill Boruki and others represent an entire class of planet never seen before the Kepler mission. As we have seen, these watery worlds probably form in the outer regions of their solar systems as ice giants similar to Uranus and Neptune, or as big-brother-versions of

Ganymede or Titan.[2] Sometime later, they may then migrate into the inner system, where their dense atmosphere and ices either condense into liquid or are stripped away by the solar wind. What's left is a planet completely blanketed in a vast ocean hundreds of km deep. But the quiet splendor of such worlds may be only skin-deep.

Herman Melville said, "When beholding the tranquil beauty and brilliancy of the ocean's skin, one forgets the tiger heart that pants beneath it; and would not willingly remember that this velvet paw but conceals a remorseless fang." The enormous pressures near the seafloor—that global ocean's "remorseless fang"—could generate unusual ices of a type not seen on Earth. This might be bad news for life, say some analysts. Because of their massive bulk and increased gravity, great pressures would build at the seafloor where mineral-rich rock meets water. Ices that form under such pressures could form a barrier between stony seafloor and ocean, effectively blocking off the carbon cycle. But scientists still have little understanding about how chemical cycling would play out on such worlds.

The geological record indicates that oceans on Earth have been present for most of its 4.54 billion year lifetime.[3] But would this longevity be shared with planets whose seas are hundreds of km deep? New studies seem to indicate that the oceans of water worlds may be long-lived affairs, lasting for billions of years.

At first blush, Earth's oceans may not seem a reliable analog for study. Although they cover 70 % of the globe's surface, they account for only one-tenth of 1 % of Earth's mass. (Like the other terrestrials, our planet is mostly rock and iron.) But that's just the cover story. The Earth's mantle—the layer beneath the outer crust—holds enough water to fill the oceans several times over. The water is continually pulled into the interior by plate tectonics. This descending water recycles back into the environment via volcanic activity, primarily at the mid-Atlantic and Pacific Ocean ridges.

Since super-Earths may have much thicker crusts than Earth does, researchers at the Harvard Center for Astrophysics decided to see if that thick shell would preclude the kind of oceanic recycling that keeps Earth's environment viable. The models investigated planets up to five times Earth's mass and up to 1.5 times its volume. The Harvard study showed that oceans on planets with 2–4 times Earth mass are even more stable than those on Earth, lasting at least

---

[2] The planet OGLE-2005-BLG-390Lb may be such a planet. It was found through microlensing methods, and may be five times as massive as Earth, but estimates portray it as an ice globe orbiting its red dwarf sun.

[3] They probably formed at the close of the Hadean Eon some 4.0 billion years ago, at a time when the surface had cooled enough to solidify and water vapor began to condense into liquid.

10 billion years. This assumes that the primary star does not swell to the red giant phase and boil the oceans away.

The biggest case studied in the model was a planet with five times Earth's mass. In this case, the thick crust staved off early volcanic activity. But once eruptions began to recharge the atmosphere, oceans became established, and once there, they were stable over the long term. In the search for life, the study suggests that older super-Earths are better targets than younger ones. In the search for extraterrestrial intelligence, the authors suggest studying planets that are a billion years older than ours. Some could argue that this says a great deal about the current status of intelligent life on Earth.

Like the hot Jupiters and hot Neptunes, water worlds may not form in place, but rather migrate in from the cooler regions of their solar systems. Heidi Hammel says that the composition of Neptune-class planets may back up this supposition. "We know that there is a lot of that material that would lead to water inside these objects. I think it's possible that that's one way you would get them. I can image another way to get water worlds: you build your solar system in an extremely rich oxygen dust cloud. Different clouds in outer space have different chemistries. If you find one that's oxygen-rich and build one there, you'll have a greater probability of having a water world. There are several different pathways that one can imagine to these different worlds."

Super-Earths in habitable zones need not all be water worlds. Studies[4] show that even if a super-Earth possesses 80 times the amount of water in Earth's inventory, the planet's surface may resemble Earth's continental and oceanic arrangement. The increased gravity of super-Earths forces the added water into the mantle, scientists claim.

Several unknowns could throw this model off. Earth's deep mantle cycling of water may not be in a steady state, which would mean that our planet's system is not as balanced as the model assumes. A second unknown is just how much water really is hidden in Earth's mantle. A third consideration is whether or not super-Earths even have plate tectonics to recycle their water. But the study's authors assert that even if these factors are off, the model is robust enough to demonstrate that water worlds with dry land masses are more frequent than ocean worlds.

If the study's conclusion is correct, it's good news for astrobiology. Continents lend stability to climate, enabling larger areas benign to life. Additionally, some biogenesis (origination of life) models infer that the

---

[4] "Water Cycling Between Ocean and Mantle: Super Earths need not be water worlds," by Nicholas Doran Cowen and Abbot, *Astrophysical Journal*, January 20, 2014.

interaction of minerals at the edge of land masses contributes richness to the chemistry necessary for life.

Some soggy super-Earths may orbit outside the habitable zone, where global seas would freeze into a deep ice crust covering the ocean beneath. Others orbit closer in, inside the habitable zones of their suns. Their oceans might hover at the boiling point, forming an amorphous transition from liquid to vapor. Seascapes on these worlds must constitute a marine hell. Sloshing waves would extend unbroken to the horizon. At their crests, boiling water would spray up into the air, turning to vapor and drifting into the glowing fog above. As the surf splashes and curls, foam would float as vast islands, accumulating bubbles from the boiling abyss. These hot water worlds would be truly alien places.

Giant Planets and Earth-Moons

The moons circling giant worlds offer a wide variety of possibilities in our search for distant Earths. Super-Jovian planets have supersized gravity fields, enabling them to harbor large moons in stable orbits. In fact, the seventh largest moon in our Solar System, Triton, orbits Neptune, which is on the smaller end of the gaseous worlds in Sol's planetary family.

Recent work reveals that moons weighing a tenth that of Earth's mass can retain a respectable, warm atmosphere, even when they orbit within the fierce radiation of a gas giant planet. Moons this size or larger could easily bear similar terrain to Earth or Mars, and even the smaller ones, if subjected to the kind of tidal heating we've seen in the moons of our own system, could have active geology with volcanoes and moving surface plates. Such small worlds may possess the same kind of feedback loops so important to life on Earth, like atmospheric recycling in the rocks and hydrological cycles.

Other Super-Earth Types

Just how "Earth-like" is a super-Earth? Those features that contribute to the uniqueness of our own world provide us with a good yardstick. First, Earth orbits in the Sun's habitable zone. Some super-Earths may well orbit at such a distance from their own stars, but studies show that this may not be enough to beget Earthlike environments. Another critical trait that enables Earth to engender life is plate tectonics. Earth's plates play an important part in our ecosystem. They recirculate the minerals that wash into the seas, and recycle elements of the atmosphere that get chemically locked into the rocks.

However, recent modelers contend that for several reasons super-Earths may not enjoy the benefits of plate tectonics. The first has to do with chemistry. It takes just the right mineral smorgasbord to create the planetary jigsaw pattern of shifting plates seen on our world. As a plate sinks, it is initially too light to descend all the way down into the mantle. But as the plate slides under another plate, a transformation takes place. At a depth of about 40 km, increasing pressures rearrange atoms within it, making its rock more dense. Without this atomic alteration, plates would stall out, ceasing to slide under each other. The change in density is dependent on the plate's makeup. And while scientists cannot yet directly measure the composition of distant Earths, they can chart elements within their stars. This data reveals an approximation of what nearby planets must be made of. Studies suggest that planets rich in silicon cannot maintain the conveyor-belt of plates seen on Earth.

Aside from composition, the crust of super-Earths may simply be too thick to carry on tectonics. Simulations of the pressures within giant Earths reveal that thick crusts likely surround most super-Earths, putting up a physical barrier to plate tectonics. In these models, the circulation of heat and minerals was too sluggish to spawn such life-critical phenomena as volcanoes. Models show that a super-Earth with ten times the mass of Earth develops a thick shell of rock unable to shift. The model planet's crust came out to about 1800 km thick, compared to Earth's roughly 25-km-thick crust. But we have never seen a super-Earth up close, and the universe is full of surprises. Some researchers assert that the increased heat within Super-Earths might be enough to drive plate tectonics after all.

Another factor that would contribute to a super-Earth's Earthliness is the presence of a magnetosphere. At the center of our world lies a molten core that generates a protective magnetic field around us. Many super-Earths probably have such an energy field surrounding them, as it is likely that they have large, molten cores.

What of those 11 billion planets more the size of Earth? The Earth Similarity Index (ESI) measures the similarity of a planet or moon compared to Earth. Its parameters include the planet's diameter, density, escape velocity (gravity) and its range of surface temperatures. The scale spans from 0 to 1, with Earth being a 1. On this scale, the ESI for Venus is 0.444, while the ESI for Mars comes in at 0.697.

Astronomers base the variables within the Earth Similarity Index on several factors, depending on the technique used for finding the planet in the first place. For example, a planet's density is influenced not only by its mass but also by its size. To date, astronomers have both mass and radius for only a small minority of confirmed exoplanets. Elisa Quintana, Kepler research scientist with NASA Ames and the SETI Research Institute, explains, "We have planets that have 5 Earth masses but a wide range of sizes. The curves are theoretical models that show the regions of mass-radius where the planets are likely made of iron, silicate rock (Earth-like) and ice, but larger ones probably have accumulated a gas envelope like Neptune or Jupiter and wouldn't have a solid surface." The take-home message, says Quintana, is that planets don't necessarily follow a predictable mass-to-radius relationship. "When we model the Kepler planets, we need mass estimates sometimes, and since Kepler only provides planet size, we (the science community) depend on mass/radius relations to give us estimates. But there is a wide diversity of planets, and we don't yet have any rule of thumb for guessing planet masses from radii (and vice versa)."

In addition to size and mass, the temperature on a planet's surface can be influenced by its surface albedo (brightness), the amount of heat falling upon it from the parent star, tidal heating and the type and structure of atmosphere (which can act as a greenhouse "blanket" to retain heat).

An ESI rating is not a direct estimate of a planet's habitability, although some of its variables influence that aspect of a planet. Worlds with high ESIs are most likely terrestrial, with rocky surfaces and sizes fairly close to that of Earth. To date, the planet with the highest known Earth Similarity Index number is Kepler 438b, checking in at 0.88 on the scale. But as for stable planets with active biospheres, we are currently left with only one example: Earth (Fig. 5.2).

Fig. 5.2 A few "giant Earths" compared. *From left to right:* Neptune, gaseous Kepler 10c, Earth, iron-clad K2-3d and the rocky behemoth BD + 20594b. (Diagram © by Michael Carroll)

Assorted Super-Earths

In 2007, data sifted by Swiss researchers uncovered a super-Earth orbiting the star Gliese 581, a red dwarf 20 light-years from Earth. In fact, three super-Earths may orbit this star. The closest to its star, Gliese 581c, orbits at the inner edge of Gliese 581's habitable zone. The team suggested that Gliese 581c orbits so close to its star that it suffers a runaway greenhouse effect much like that found on Venus. The other two, Gliese 581d and Gliese 581g, may orbit out in the habitable zone, making them candidates for life. But the planets serve a cautionary tale as to how difficult these measurements are. Just after the announcement, other researchers called into question the very existence of the two planets. Follow-on analysis, however, seems to have confirmed the existence of this remarkable "three-Earth system."

The outermost of the trio, Gliese 581g, orbits just 0.13 AU from the star. But because of the red dwarf's dim nature, the planet receives roughly the same amount of energy as Earth does from the Sun. It probably has a radius no larger than 1.5 times that of Earth, barely qualifying it for super-Earth status. All of these factors combine to make Gliese 581g one of the most—if not *the* most—of Earthlike worlds beyond our own. Its Earth Similarity Index may be as high as 0.9.

Researchers have been able to fill in a more detailed portrait of Gliese 581g, thanks to some sophisticated computer modeling. The planet is close enough to its sun to be tidally locked, always keeping the same face toward the star. Depending on the composition of its atmosphere and surface, the planet may be a barren, Venus-like world, but if conditions are right, and it is farther out in the habitable zone, Earth's big brother may be quite different. With atmospheric pressures similar to Earth's, models indicate that the globe may be blanketed in a thick ice crust. Without the effects of an atmosphere, global temperatures may range from –64 °C to –45 °C. But if the air contains enough greenhouse gases, including carbon dioxide, temperatures may be substantially warmer.

Because of the fact that it is tidally locked, Gliese 581g's atmospheric circulation may lead to a permanent ocean facing the star, just at the substellar point (with the sun directly overhead). Temperatures in this region would be balmy, similar to those in Earth's tropics. Circulation studies indicate that wind patterns would lead to a great, sideways-pointing chevron-shaped warm region. Gliese 581g shows us that it is theoretically possible for a tidally locked planet to support life. If its distance and temperatures are similar to those of Gliese 581g, or if it has an atmosphere that creates stable environments along

Fig. 5.3 The planet Gliese 581g may be the most Earthlike world in the galaxy. Its tightly wound orbit around its red dwarf sun provides conditions favorable for life. Models indicate that under the right conditions, a chevron-shaped permanent ocean would spread across the super-Earth at the substellar point. (Art © Michael Carroll)

the terminator (the day/night border), conditions may be favorable for life even on such an alien world.

Another world in the system, 581d, is significantly more massive than Earth, at roughly 6.8 Earth masses. The huge size of the planet—five times that of a super-Earth class—caused astronomers to add a new class to exoplanets, the mega-Earth. When first discovered, estimates put the planet at the outer edge of the habitable zone, but recalculations in April of 2009 show that it orbits closer in, with a period of 66.87 days. This places the planet at a distance where liquid water can exist (Fig. 5.3).

The nearby star system Gliese 667 (or GJ 667) contains three stars some 22 light-years from Earth. One is a red dwarf, while the others are K-type orange dwarfs slightly cooler than Sol. Around the K star Gliese 667C, we find at least three planets. This is surprising. Orange dwarfs like Gliese 667's larger couplet are somewhat depleted in the kinds of heavy elements that are the brick-and-mortar of terrestrial planets. Since these stars contain only 25 % as much of this material as our Sun does, astronomers thought it unlikely that star systems like Gliese 667 would possess low-mass rocky planets. But like so much of exoplanet research, the triple-star system broke the mold. It may harbor three super-Earths close to—or within—its habitable zone.

The planet Gliese 667Cc has a mass approximately 1.5 times that of Earth. The alien world may be a rocky terrestrial, making it one of the most Earthlike planets known, but most estimates put the planet more in the nature of a dwarf Neptune. Like Kepler 69b, GJ 667Cc circles its sun at a furious speed, completing a circuit in just 28 days. But GJ 667C is a red dwarf star, so although the planet's orbit is tight, it is far enough out for liquid water to exist on the surface. Gliese 667Cc collects about 90 % the light that Earth receives from the Sun. While it is in the habitable zone, it is still within reach of the red dwarf's energetic solar flares, which may periodically sauté the surface in deadly radiation. The oversized Earth is so close as to be tidally locked to its parent star. GJ 677Cc has spectacular skies. Its red dwarf is a member of a triple-star system. Gliese 667 A and B are both K-type (orange dwarf) stars in orbit around each other. They complete their dance once each 42 years, at a distance ranging from 5 to 20 AU. Gliese 667C orbits them both at a distance of 230 AU, nearly eight times the distance between Neptune and the Sun.

Planet GJ 667Cc may have plenty of company. A sub-Neptune, GJ 667Cb resides at less than half the distance of GJ 667Cc's orbit, making each circuit of its sun in a little over 7 days. The planet is at least 5.6 times as massive as Earth. Three additional planets were originally claimed, but these have been called into question because of noise in the data caused by fluctuations in the star itself. Confirmation of these planetary siblings awaits additional observations (Fig. 5.4).

At a distance of 600 light-years away,[5] the Sun-like star Kepler 22 husbands a family of planets that includes the super-Earth Kepler 22b. The planet has the distinction of being the first planet orbiting within its parent star's habitable zone discovered by the Kepler Observatory. Its density, similar to that of rock, means that the world's nature may be terrestrial. Kepler 22b is larger than Earth, weighing in at 2.5 times the mass. It may have a denser atmosphere, and as it orbits in the inner region of its star's habitable zone, its climate may resemble Venus more than Earth, but other factors (rotation, cloud cover) might moderate temperatures there, resulting in a big brother to Earth. Some models now point to surface temperatures hovering around a comfortable 22 °C (72° F). The planet is large enough that, rather than terrestrial, it may be a sub-Neptune, with a deep, cloud-banded atmosphere.

---

[5] At its distance of 600 light-years from here, any astronomer on Kepler 22, viewing our planet today, would see Earth as it was in the Renaissance.

Fig. 5.4 Seen from a nearby moon, the super-Earth Gliese 667Cc may be a sub-Neptune, with windy cloudscapes rather than rocky vistas. The planet is so close to its orange dwarf star that it is probably tidally locked, a situation that may wreak havoc with its banded cloud formations. Yellowish clouds of sulfur tint some areas green. (Art © Michael Carroll)

In 2012, astronomers at the European Southern Observatory announced the detection of a planet orbiting the nearby star HD 40307. Using the radial velocity technique, the researchers found the super-Earth orbiting within its star's habitable zone. HD 40307 lies just 42 light-years from Earth, close enough that future observatories may well be able to resolve its super-Earth's spectrum in search of life signs. The planet orbits some 90 million km from its orange dwarf star, half the distance between Earth and Sun. It is large enough that its composition may range from a large terrestrial to a sub-Neptunian with no firm surface.

Investigators using the High Accuracy Radial Velocity Planet Searcher (HARPS) facility have announced the discovery of 49 planets. Among them is the super-Earth HD 85512b. With 3.6 times the mass of Earth, this rocky super-Earth orbits the orange dwarf star Gliese 370, some 36 light-years from here. The planet is approximately within the inner edge of the habitable zone.

Tau Ceti e and f are two super-Earths circling the star Tau Ceti, a star similar to the Sun, but smaller and slightly dimmer. Tau Ceti e circles close to the star in the hot region of the habitable zone, with conditions that may approach those of Venus. Tau Ceti f orbits farther out, close to the outer edge of Tau Ceti's habitable zone. Tau Ceti f is 6.5 times the mass of Earth, and may accommodate conditions permitting liquid water, especially if its atmosphere presents a substantial greenhouse effect to warm its environment.

Two other exo-Earths in the size range of ocean worlds orbit within a habitable zone of the orange dwarf Kepler 62. Both are roughly half again as large as Earth, putting them at the border between Earthlike and super-Earth. Each is a possible candidate for having a rocky, terrestrial surface. With the right combination of atmospheric gases and cloud cover, both worlds could sustain liquid water on their surfaces. Kepler 62f (ESI rating 0.67) may have an atmosphere denser than Earth's, perhaps similar to, but cooler than, that of Venus. In fact, studies indicate that Kepler 62e, with an ESI of 0.83, is likely a water world, swathed by a global, deep ocean. Sibling Kepler 62f may also have a large component of water, but may be far enough out in the habitable zone to have congealed a frozen surface, at least at its poles. Barclay points out that "Water is very common. If you've got the right temperature, you're going to be full of water. [Kepler 62e or f] may well have the right temperature; it's probably too big to be a typical rocky planet like ours. You'd probably have some hydrogen and helium, but you're going to have oxygen, so there's going to be water. These are not like our oceans. We're talking hundreds of kilometers deep." Many more super-Earths may be water worlds, but the error bars in the data are still fairly wide, so nailing down specific conditions on such alien worlds awaits more information (Fig. 5.5).

One of the closest Earth-sized planets lies 39 light-years away. Circling the dim red star GJ1132 (Gliese 1132), the rocky exoplanet is 1.2 times the mass of Earth. GJ1132b likely orbits closer to its sun, too close to retain liquid water on its surface. It orbits once each 1.6 days and was discovered by the MEarth-South Observatory in Chile. Because the nearby planet crosses the face of its star when viewed from Earth, telescopes akin to the HST could theoretically tease out the composition of the atmosphere by observing the starlight passing through it. GJ1132b may well be the beginning of a new chapter in our study of Earthlike worlds, offering the chance to actively search for signs of life within the gases of distant atmospheres.

What's the Forecast?

The varied atmospheres of exoplanets remain difficult to study, but with advances in technology, today it is possible to examine a planet's atmosphere using spectroscopy while it is transiting. When the planet transits the star, light from the star passes through the upper atmosphere of the planet.

Fig. 5.5 The exo-Earths Kepler 62e (*top*) and 62f may both be water worlds. 62e orbits its red dwarf sun at a distance similar to the distance at which Mercury orbits the Sun, but conditions there permit surface water. Methane, chlorine and other gases may contribute to a sky of a different color than our own. Kepler 62f orbits farther out, and may display vast reaches of frozen surface water, remnants of an ancient global sea. (Art © Michael Carroll)

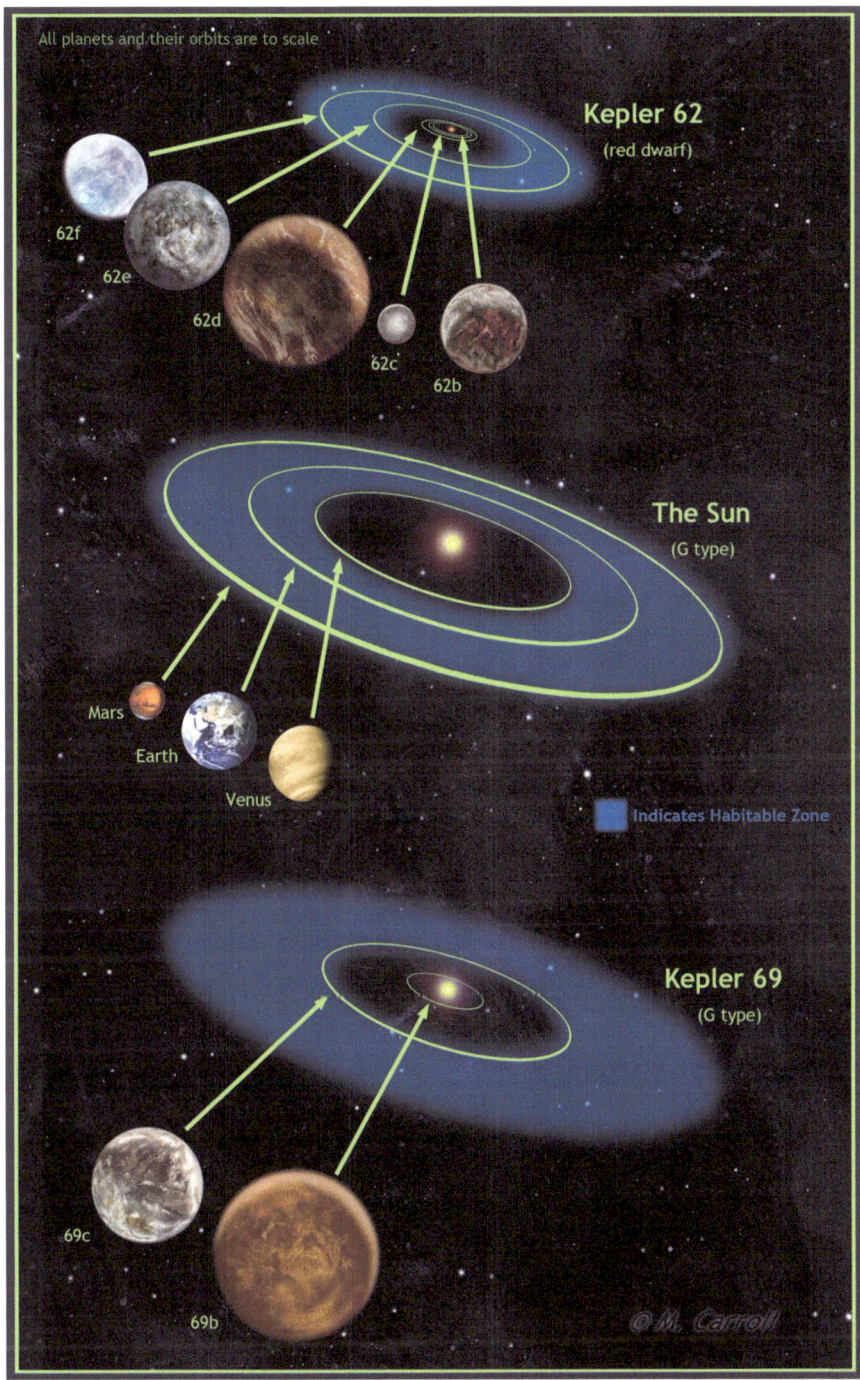

Fig. 5.6   The Earthlike planets of Kepler 62 and Kepler 69, compared to our own system's inner planets. (Art © M. Carroll)

By studying the star's spectrum carefully, astronomers can tell the makeup of the planet's atmosphere. A planetary atmosphere could also be detected by measuring the polarization of the starlight as it passes through—or is reflected by—the planet's atmosphere. This technique has led to some fascinating insights, says NASA's Mark Marley. "There's a lot of data from the planets that transit. People have put a lot of effort into measuring the spectra. They've seen water and other gases. Some of the planets appear to have clouds or hazes that are making their atmospheres very opaque. This is still a big topic of research into what exactly is going on. When you look at the sunset around the planet, why is it such a sharp cutoff at every wavelength? Maybe there is a cloud or haze layer that's causing it."

On the far side of a planet's orbit, when the planet is blocked by its star, observers can directly measure the planet's radiation. If the star's brightness during the planet's secondary eclipse is subtracted from its intensity before or after, only the signal caused by the planet remains. Researchers can then measure the planet's temperature and even discover possible signs of cloud formations on it. This information may even include indicators of life, such as the presence of free O_2 in the atmosphere.

As for atmospheres, spectral studies may one day tell us what kinds of gases enfold Earthlike planets. But for now, we use models to estimate what conditions may be like on a given exo-Earth. It's not an easy task, warns Thomas Barclay of NASA Ames. Barclay is part of the Kepler planet-hunting team, and dedicates much of his time to characterizing what conditions may exist on Kepler's discovered worlds. "These models are incredibly simplistic. We're just reaching good models of our own weather, but they're not great yet, so trying to do that on an exoplanet is highly speculative. These are more guidelines than hard and fast rules. Even the people whose careers are doing these models will say that the point of defining a habitable zone isn't to understand what an individual planet is like. It's more like being able to create a target list, a ranking. If we've got a limited amount of resources, which is the best planet to spend our resources on? For that, it's pretty useful."

In February of 2016, the Hubble Space Telescope was able to detect hydrogen, helium and perhaps hydrogen cyanide in the atmosphere of 55 Cancri e, a super-Earth twice the diameter of Earth. This marked the first detection of specific gases in the atmosphere of a super-Earth. With a density of 8.63 that of Earth, 55 Cancri e may consist of nearly pure carbon, an alien world indeed.

Exoplanet meteorologists continue to make progress on the weather front. For the first time, a team of researchers at the University of Warwick has directly monitored weather systems on a planet beyond our own Solar System.

Fig. 5.7 Furious winds fan out across the searing face of hot-Jupiter HD 189733b, a giant in a two-star system. The planet's 53-h orbit keeps it tidally locked to its orange dwarf primary star. An M dwarf star is behind to the right. (Art © Michael Carroll)

The weather report comes from the planet HD 189733b, orbiting the primary star in a binary star system 63 light-years away. The primary star in the HD 189773 system is an orange dwarf, slightly smaller and cooler than Sol. The second star is an M star. HD 189773b orbits the primary once every two days. It is a hot Jupiter, 13 % larger than Jupiter. If a planet's Earth Similarity Index could be in the negative, this planet would be there. Temperatures at its cloud tops probably reach 1000 °C (Fig. 5.7).

Researchers have clocked winds at over 2 km/s, a speed twenty times greater than Earth's hurricane-force gales, traveling seven times the speed of sound. Although winds have been found in exoplanet data before, this was the first direct measurement of flow. Astronomers were able to measure wind speed on two sides of the fast-moving planet, resolving currents of 8690 km/h from the day-lit hemisphere to the night side. The team utilized high-resolution spectroscopy from sodium in the planet's atmosphere. The moving atmosphere has a Doppler shift, in similar fashion to a star's movement in the radial velocity technique. As the planet transited from the star's edge to its brighter center, the amount of light blocked by the planet's atmosphere changed. This change could be used to measure the wind velocity on opposite sides of the planet, revealing the wind speeds.

The first multiple planetary system to be directly imaged around another star also yielded insights into exoplanet atmospheres. The system was a family of globes orbiting the star HR 8799. A Jupiter-like planet orbits at a distant 68 AU from its star, taking 460 years to make the circuit. HR 8799 b's radius is about 30 % larger than Jupiter's, and its mass is estimated at between 4 and 7 Jupiter masses. It is the outermost planet in HR 8799's system. Astronomers discovered the planet, along with its two siblings, using the Keck and Gemini observatories in Hawaii. All three planets were discovered using the direct imaging technique.

Photometry—the study of light—of HR 8799b shows that it has thicker clouds in its atmosphere than do planets with higher gravity but the same temperature. Near infrared imaging of HR8799b revealed a dusty atmosphere rich in hydrogen. Further studies carried out at Palomar Observatory show indications of ammonia, acetylene, carbon dioxide, and traces of methane.

The case of HR 8799b was aided by the massive size of the planet. But as researchers hone their skills, Earthlike planets will soon yield their own detailed weather reports.

Zeroing in on Earth 2.0

The popular press has touted many exo-Earths as being "Earth's twin," but only a handful of planets come close to fitting the bill. If we survey Earthlike alien planets using the Earth Similarity Index, we come up with even fewer. With the ESI of Mars standing at 0.697 and that of Venus at 0.444, we get a better idea of how close to terrestrial conditions these exoplanets come.

One of the most Earthlike planets yet discovered is a world only 12 % larger than our own. Kepler 438b (also designated KOI-3284.01) clocks in with an Earth Similarity Index number of 0.88, the highest of any known. It orbits within the habitable zone of a red dwarf, circling the star every 35 days. If Kepler 438b is terrestrial in nature, its mass will be 1.4 times that of Earth. Surface temperatures on the distant Earth will range from 0 °C to 60 °C (from the freezing point of water up to 140°F).

The planet has the disadvantage of orbiting close enough to its parent star that it can feel the fallout of solar flares so common to red dwarfs. In fact, Kepler 438 is very active, unleashing flares of radiation and plasma, called coronal ejections, every few hundred days. Recent work by scientists at the Harvard-Smithsonian Center for Astrophysics paints a harrowing picture of the environment around a red dwarf. The flow of charged particles from the star, its "solar wind," is strong enough to strip away the atmosphere of a close planet, and if a planet orbiting a red dwarf is to have liquid water, its orbit will be close, some 15–30 million km. (Remember that Mercury orbits Sol at a distance of 58 million km). But the pressure of the solar radiation is about the same for sun-like G stars and red dwarfs. Red dwarfs also have a more energetic magnetic field, and they emit more X-ray and UV radiation, which create greater erosion on a planet's atmosphere. But as for Kepler 438b, atmospheric ozone and a strong magnetic field from a molten rocky core could shelter the planet's surface, allowing life to thrive even there (Fig. 5.8).

A second close match to an Earthlike planet is Kepler 452b, the first roughly Earth-size planet to be found in the habitable zone of a star similar to the Sun. With an ESI rating as high as 0.83, the planet is approximately half again as large as Earth. Although Kepler 452b is slightly farther from its star than Earth is from Sol, Kepler 452 is brighter, so the planet gets just a bit more energy from its star than Earth does from the Sun. Any terrestrial vegetation transported there would thrive in similar lighting conditions and temperatures. That is, *if* Kepler 452b has a solid surface. One recent study gives the planet a 13 % chance of being terrestrial rather than gaseous. The planet's size hovers right on the edge between a rocky super-Earth and a sub-Neptune. If Kepler 452b is terrestrial in nature, its atmosphere is probably thicker than Earth's. A rocky world that size may well have active volcanoes. On Earth, most volcanoes erupt at the boundaries of its drifting continental plates. Although a super-Earth may have a crust too thick to support plate tectonics, its larger bulk would retain more internal heat, perhaps leading to localized mantle plumes that would result in volcanic eruptions (Fig. 5.9).[6]

It takes 385 days for Kepler 452b to orbit its sun, a year quite similar to Earth's 365-day year. But all may not be well on Kepler 452b. Its star, older than the Sun by 1.5 billion years, is reaching a more active phase in its development, sending out more heat and light than it previously did. Kepler 452b was once in the center of its star's habitable zone, but as the parent star has aged and temperatures have risen, its habitable zone has migrated out, stranding

---

[6] For more on the subject, see "Geodynamics and Rate of Volcanism on Massive Earth-Like Planets" by E. S. Kite, et al; *Astrophysical Journal,* 700: 1732–1749, August 1, 2009.

Fig. 5.8 Born under an angry star, Kepler 438b looks out beyond the clouds to a red dwarf sun. The planet is near enough to enjoy the sight of the star's massive flares. Unfortunately, it's also close enough to suffer from deadly radiation. (Art © Michael Carroll)

Fig. 5.9 Super-Earth Kepler 452b may be a sub-Neptune, but some estimates suggest it is rocky, and large enough to have retained much of the heat from its creation, heat that might express itself as massive volcanoes. The planet may have had vast oceans at one time, but its star is warmer now, so surface water may be evaporating on a global scale. (Art © Michael Carroll)

this planet on the inner edge. Any oceans it once had are evaporating into its thick atmosphere. Venus may have followed a similar course of development. Early in our own Sun's life, the star was cooler, so its habitable zone was closer in. Today, Sol's HZ has migrated out, and both Venus and Earth are warmer than they once were.

Super-Habitable Places

Some astrobiologists are more optimistic about the frequency and extent of habitable zones. It has even been suggested that some planets may be more habitable than Earth. Researchers call these "super-habitable" planets. Instead of focusing on terrestrial planets or moons within habitable zones, researchers are now considering a wide range of properties that could contribute to habitability. Tidal heating, for example, can make up for a frigid environment by adding heat to the planet's potential biome. The gravitational push and pull of nearby moons generates the kind of internal heat that could enable life to thrive. Jupiter's volcanic moon Io provides a typical example. This chilly satellite is about the size of Earth's Moon, and was assumed to be geologically dead. But the gravitational tug from nearby Jupiter and the moons Europa and Ganymede trigger inner heating (see Chap. 6).

Another factor leading to habitability may be the surface area and nature of a planet. More surface area affords more room for diverse biomes. Earthlike worlds with similar land surface to Earth, but with continents broken up into smaller provinces, may be more habitable with their plethora of shorelines. Large continental regions (like our own supercontinent Gondwanaland, which existed some 500 million years ago) may have uninhabitable wastelands within their interiors. The shallow waters of such worlds, like those on Earth, may have greater biodiversity than deep oceans, gifting these super-habitable worlds with even more life-friendly real estate than that on Earth. But even desert worlds may have an advantage: watery worlds are high in humidity, and that humidity may contribute to superheated atmospheres. But desert regions, with their low amounts of water vapor, may actually be cooler overall. If lakes and oases scatter across these types of planets, they may be super-habitable as well. Increased oxygen levels in atmospheres could facilitate life forms even larger than the dinosaurs, and denser atmosphere can shield the surface of planets that orbit near stars. Some estimates put Earth at the very inner edge of Sol's habitable zone. This means that planets in the center of a given zone will be more habitable as well.

One super-habitable world may orbit a member of the closest star system to Earth, Alpha Centauri. Alpha Centauri B, an orange dwarf, is cooler and older than the Sun. Astrobiologists estimate that super-Earths with masses 2–3 times that of Earth might be ideal candidates for life. Any planetary system orbiting Alpha Cantauri B will be older than our own, with an age of from 4.8 to 6.5 billion years. If life on a super-habitable planet of Alpha Centauri B followed a similar path to that of terrestrial life, microbial colonies may have been thriving on its planet back in the days when our home world still suffered under a hail of asteroids and comets, and reeled from the impact that led to the Moon. In fact, astronomers at the European Southern Observatory have tentatively identified an Earthlike world orbiting nearby Proxima Centauri.

Before the discovery of Kepler 452b, exoplanet experts considered a rocky world called Kepler-186f to be the most Earthlike one found. 186f is a mere 10 % more massive than Earth. The planet gets only about a third of the energy that Earth does, as its M dwarf star—large as dwarfs go—is half the mass of the Sun. At the height of day, surface light would be only as bright as Earth's an hour before sunset.

Kepler 186f has four confirmed siblings, all with probable solid surfaces. They range in mass from 8 % larger than Earth to 40 %, qualifying all as Earthlike or super-Earths. Most orbit close enough to their star that they are tidally locked, but 186f puts enough distance between itself and its sun to have an independent day/night cycle.

"When we found it, we could tell that it was clearly something interesting," says NASA/Ames' Tom Barclay. "At those early stages, you're never quite sure whether it's going to pan out or not, but this one really did."

Confirmation took a painstaking year of hard work to really understand the complex cotillion of planets in the Kepler 186 system. Elisa Quintana outlined the discovery. "Kepler 186, the star, actually had four planets detected early on. They are all roughly Earth-sized, less than 1½ Earth radius. The inner four have orbits between 3 and 22 days. They're also so close—and the star comparatively small—that they block a larger proportion of the light. Kepler 186 is an M dwarf, about half the size and mass of the Sun. Its habitable zone is close in, so planets within the HZ will pass in front of their star frequently. [Their periods are so short that they repeat quickly, making them easier to detect.] The more transits you have, the stronger the signal. But Kepler 186f orbits with a 130-day orbital period, so it took a lot more data to see that planet transit three times."

The researchers were cautious about releasing the news until they were sure. They had been burned before, Quintana says. "One of the first times we got really excited was with Kepler 69. Originally it looked like a true Earthlike planet around a sun-like star. They popped open the champagne and we celebrated, but Tom (Barclay) spent the entire weekend modeling the star. It turned out that the star was slightly bigger than we thought, so the planet is more like Venus. It was a good cautionary tale."

Although Kepler 186f looked like a nice, Earthlike planet within its star's habitable zone, researchers spent a year nailing down the size of the star. The star's remote location, some 500 light-years away, didn't help matters. After getting some ground-based data, the team was able to validate the size of the star and, hence, the planet's characteristics, and Kepler 186f remained Earth-sized. "With Kepler," Quintana says, "the main goal is to find a true Earth. There were these milestones along the way, including the first Earth-sized planets, Kepler 20e and f. Kepler 22b was the first habitable zone planet, but it was almost twice the radius of Earth. Now we know that that's too large to have a rocky surface. 186f was the first definitive Earth-sized planet found around another star with a solid surface in the habitable zone. It was Kepler's proof of concept."

What would an interstellar traveler see from the surface of such a world? The sun in the sky would take on a distinct peach tinge, says Tom Barclay. "Standing on Kepler 186f, the sun isn't going to be some glowing red orb like the eye of Sauron. It's just tinted toward the warmer colors. Still, I think the sunsets would probably be pretty beautiful." The star's apparent diameter would be $1^1/_3$ the size of the Sun in our own sky. While the star itself is only half as far across, the planet is less than a third as far away as Earth is from Sol (in fact, the entire Kepler 186 planetary system would fit inside the orbit of Mercury). At high noon, light levels would be equivalent to those on Earth an hour before sunset.

As is the case with many other exoplanets, error bars for surface conditions are fairly wide. If its orbital distance is on the lower end of estimates, the planet might bask in global desert conditions, perhaps with extensive cloud cover from water vapor. Farther out, the world might see more liquid surface water, with lakes, oceans, waterfalls and rainstorms. At the far edge of Kepler 186's habitable zone, less likely, the planet's water inventory might be frozen into glaciers and vast snowy wildernesses, peppered here and there by lakes or rivers. Kepler 186f is big enough to have mountain ranges, canyons, and perhaps volcanic or other tectonic activity. Waterfalls may stream from its highlands, perhaps feeding into lakes and seas. And if things have gone just right for the planet, its slopes and valleys may even be blanketed by vegetation (Fig. 5.10).

The planet that provided Barclay's cautionary tale, Kepler 69c, is another addition to the super-Earth parade. It revolves around the star Kepler 69, a G-type star very similar to the Sun. Kepler 69 is 93 % the size of the Sun and 80 % as bright. Its attendant, Kepler 69c, circles its parent star every 242.5 days. This means that the planet receives about as much energy as Venus does. Kepler 69c may, in fact, be more of a super-Venus than a super-Earth. Kepler 69c also has a brother: Kepler 69b is over twice the size of Earth, and orbits its sun in a breakneck 13-day loop, making surface temperatures hotter than those at Mercury.

Kepler 69c has 70 % more mass than Earth does, so is another borderline Earthlike planet. The distance from the planet to its star is half that between Earth and Sun. Its 242 day, orbit is comparable to Venus in our system. But while Kepler 69 is a G-type star, like the Sun, it is slightly dimmer—80 % as bright. This puts the planet at the inner edge of the star's habitable zone. The surface of Kepler 69c may be rocky—similar to the terrestrials in our Solar System—or it may be oceanic. The planet may be far enough into the habit-

Fig. 5.10   The M dwarf star Kepler 186 stares down upon a benign environment in this artist's view of planet Kepler 186f. Temperatures may be quite Earthlike on this cousin of our home world, with liquid water and a rocky surface. The pinkish sun in the sky would appear nearly half again the size of the Sun in Earth's sky. (Art © Michael Carroll)

able zone for an Earthlike environment with continents, lakes, seas and oceans, but most researchers estimate that its temperatures likely result in Venus-like conditions with a dense, hot atmosphere. It is the smallest Earthlike planet yet found within a star's habitable zone.

Decoding the data to understand exoplanet conditions is a difficult process. "There are now ~60 confirmed planets with radii of less than 4 Earth radii for which we have both mass and densities," says Elisa Quintana. "There are 26 with radii less than 2 radii of Earth that have both mass and radius. So there are less than two dozen where we are fairly sure of their composition." In other words, 26 of the 60 are probably rocky, while the others will trend toward gaseous sub-Neptunes. The K2 mission hopes to add 75 planets to this number.

"There's a number of really nice candidates out there," Barclay says. "Each one of them that we've found is different enough from Earth in one or more regions to not quite be an Earth analog. 186f, for example, is somewhat Earthlike in size, but orbits a different type of star. We're still honing in on something that we can say, 'That's like where we live.'"

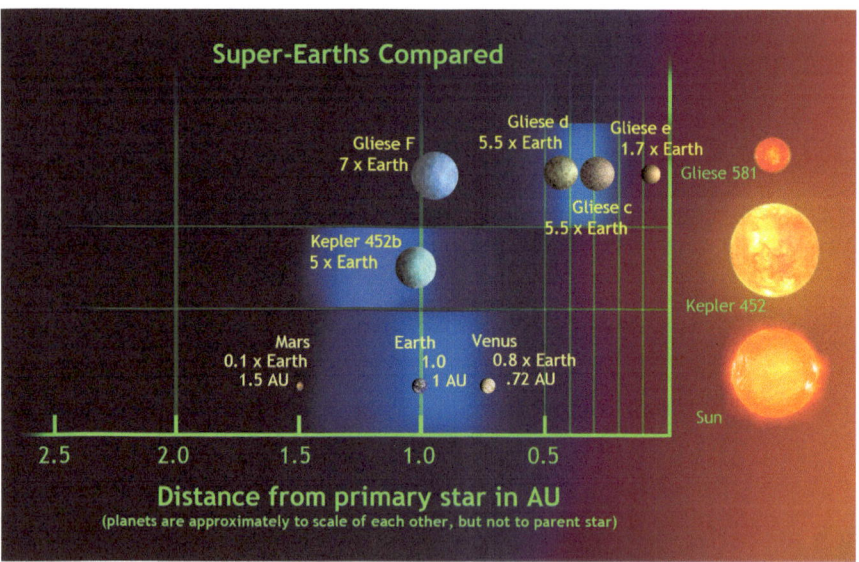

Fig. 5.11 A tale of three star systems: the tentatively-identified super-Earths of Gliese 581 (a red dwarf) and Kepler 452 (slightly hotter than the Sun), compared to our own Solar System. Blue regions indicate approximate habitable zones. Note that some astronomers consider the Sun's habitable zone stretching out as far as the orbit of Mars. Planets are not to scale. (Diagram © Michael Carroll)

And although observations are teaching exoplanet researchers about trends, the details of conditions on these distant worlds are elusive (Fig. 5.11). As Elisa Quintana points out, "Every once in a while a planet ruins what people think."

Among the other assorted Earth contenders, Kepler 442b is likely a solid-surfaced terrestrial world less than 1.5 times the mass of Earth. It is a third larger, and orbits its M dwarf star once every 112 days, putting it in Kepler 442's habitable zone. Super-Earth Kepler 443b orbits an orange dwarf star. The planet circles toward the inner edge of its star's habitable zone. Kepler 443b is large enough (2.3 Earth diameters), and with a low enough density, to qualify as a sub-Neptune rather than a terrestrial.

Twin worlds may inhabit the HZ of the double-star system Kepler 296. When Kepler 296f was first discovered, its sun was thought to be a loner. But further studies by the Hubble Space telescope revealed that Kepler 296 is actually a double star, with both members being M dwarfs: Kepler 296e and f. New analysis shows that both Kepler 296e and Kepler 296f orbit in the habitable zone. Kepler 296e is bordering between rocky and gaseous, so its "Earthlikeness" is in question. Additionally, the planet orbits at the inner edge of the HZ. It is probable that Kepler 296e has conditions similar to a cool Venus. Kepler 296f is more promising as a biologically amenable super-Earth. It is larger, so it is probably either a water world or sub-Neptune, but it is well within the habitable zone. If the planet has any large moons, these may be candidates for Earthlike worlds. Astronomers are in the process of confirming many more planets in the super-Earth range.

Giants Inside the Habitable Zone: The Promise of the Moons

A swirling host of moons surround all four of the giant worlds in our own Solar System. Some are as complex—and as sizable—as planets. But within the life-friendly habitable zones of other stars, it is the gas giants that make up the majority of the known planets. According to a recent study,[7] "the moons of giant planets could actually be the most numerous populations of habitable worlds." Three large moons in our own system are considered candidates for life or life's precursors: Jupiter's Europa, Saturn's Enceladus, and Saturn's Titan (see Chap. 6). If we took these moons and brought them into the HZ of Sol's system, they would become oceanic worlds filled with the promise of

---

[7] Heller, R., et al. 2014. "Formation, Habitability, and Detection of Extrasolar Moons." *AsBio*, 14, pp. 798–835.

Earthlike environments. Since many giant exoplanets circle their stars within the habitable zone, they provide places to search for exo-Earths tagging along. No moons of exoplanets have yet been detected, but researchers continue to pursue new techniques for finding such small worlds orbiting giant ones.

Typical of the giant worlds orbiting within habitable zones is Kepler 47c. What's not typical of the planet is that it orbits two suns. The Kepler 47 system is binary, with the primary star similar to our Sun, but 84% as bright. The second star is a small, M type a third the size of the Sun. Kepler 47c orbits within the habitable zone encircling both stars. But the shared habitable zone of this two-star system is complex. As the primary star circles around a point between itself and the smaller star once every 7½ days, its HZ precesses, wobbling around the star. Kepler 47c appears to remain within this varying zone. Its year spent circling the stellar duo lasts 303 days. The planet itself is a super-Neptune, gaseous and probably not hospitable for life. However, any moons around the planet, if large enough to hold atmosphere and water, may well be Earthlike.

Another planet in the system, Kepler 47b, orbits closer in to the stars. Its 50-day circuit keeps conditions on its surface roasting, with an atmosphere that may be more Venusian than Earthly.

Another gaseous goliath orbits within the habitable zone of the binary star system Kepler 16. The primary star is an orange dwarf slightly smaller than the Sun. It orbits around a point between itself and a red dwarf. Like Kepler 47b, planet Kepler 16b orbits both of the stars in their shared habitable zone, making the yearly trek once each 229 days. The planet is similar to Saturn in mass, with an estimated composition of half gas and half rock and ice.[8] Kepler 16's habitable zone is estimated to stretch from 55 to 106 million km from the binary stars. Kepler 16b orbits in the zone's outer region, at about 104 million km distance. The planet should have temperatures ranging from −100 °C to −70 °C. But although ambient temperatures are cold, the right mix of greenhouse gases could keep the surface of Kepler 16b above the freezing point of water. As for Earthlike companions, computer models indicate that sometime in the system's history, other planets may have perturbed an Earth-sized inner planet from another location in the habitable zone, causing it to migrate out of its orbit, where Kepler-16b eventually captured it as its moon. If so, Kepler 16b may be one of many worlds with Earthlike moons in tow.

---

[8] Because the planet was caught transiting two stars, more data was available than it would have been in a single-star system. The planet's mass is known to a precision of 0.3%, a value more accurate than for any other exoplanet, as of September of 2011.

6

Looking for Life in All the Right Places

Life Based on What?

For the miners on the asteroid Janus VI, life has gotten uncomfortable. Their operations have been disrupted by a life form based on silicon, a creature that lives in—and travels through—solid rock. After several terrifying scenes of technicians being mauled by the mystery monster, we finally see the alien. Alas, it appears to be a guy under a carpet with orange squiggles drizzled across it. But while lacking in a special effects budget, *Star Trek*'s episode "Devil in the Dark"[1] reveals tremendous creativity in terms of imagining alien life. The Horta, it turns out, is silicon-based, intelligent, and kind (once the bumbling humanoids are able to talk to it). Could sentient beings live in completely alien environments? Could they, in fact, be based on something besides carbon?

Life in Our Own Image

We are limited in our definitions of life and of intelligence. We have only one example on our planet. This geocentrism provides a barrier to our understanding of the cosmos and what may inhabit it. As science fiction writer Stanislaw Lem puts it, "Modesty forbids us to say so, but there are times when we think pretty well of ourselves. And yet, if we examine it more closely, our

---

[1] *Star Trek* original series, season 1, episode 25, written by Gene L. Coon.

© Springer International Publishing Switzerland 2017
M. Carroll, *Earths of Distant Suns*,
DOI 10.1007/978-3-319-43964-8_6

enthusiasm turns out to be all a sham. We don't want to conquer the cosmos…
we simply want to extend other boundaries of Earth to the frontiers of the
cosmos. For us, such and such a planet is as arid as the Sahara, another as
frozen as the North Pole… We are only seeking Man…we go in quest of a
planet, a civilization superior to our own but developed on the basis of a pro-
totype of our primeval past."[2]

Astrobiologist Chris McKay cautions that geocentrism has already affected
our search for life among the nearby planets. McKay refers to the phenomenon
as confirmation bias, "which is a fancy way of saying you tend to see what you
want to see." McKay cites cancer research as one of many examples.

> People will do a trial run and pull out a certain result, but the result is sort of
> what they were hoping for, so they glom onto it. Astrobiology suffers from con-
> firmation bias. "I found life in a meteorite!" or "I found life in the red rain that
> I collected in the stratosphere, and I don't know what it is, so it must be extrater-
> restrial life." That is the default explanation. "If I can't prove that it's not extra-
> terrestrial life, then it must be extraterrestrial life." You just have to take a very
> rigorous view of the data. As a scientist, you're trained to say "Here's the data;
> here are the explanations consistent with it; here are the problems with those
> explanations," rather than taking an advocacy stance.

In McKay's opinion, astrobiology attracts people—usually from other
fields—who are so enthused about the search for extraterrestrial life that
they set aside critical thinking. Rather than objective observers, they become
advocates.

> One of the most recent cases was one where they were convinced that buried
> inside of this meteorite was a cyanobacteria. They said "It just looks so much
> like a cyanobacteria that it's just got to be a cyanobacteria." I set aside that and
> said, "Okay, suppose it looks like a cyanobacteria. A cyanobacteria needs sun-
> light and it needs water and, at least on Earth, they all use oxygen." It becomes
> so improbable that what you're seeing is, in fact, a cyanobacteria. It came from
> the interior of an asteroid. Even if this thing was, at some point, on the surface
> of this asteroid, you have to invoke some kind of atmosphere so this thing
> could live at the surface with some kind of liquid water. It's not impossible, but
> it strains our understanding of things to the point that I'm willing to say this looks
> like a cyanobacteria, but it's not. It's a real problem, and it goes into exoplanets
> as well.

---

[2] From his novel *Solaris*; see Chap. 7.

Knowing whether an environment can support life is different than actually recognizing living systems. The difficulty in identifying alien life forms is underscored by our difficulty in recognizing life forms here at home. Whenever researchers make a claim pertaining to astrobiology, McKay says the scientific community must be circumspect. They got their chance with a meteorite from Mars.

A Martian Invasion of a Different Sort

At the end of the Transantarctic mountain chain lies a 12-km stretch of unassuming, low-lying hills. Resting at the head of two Antarctic glaciers, the Allan Hills form a natural barrier to the flowing glacial ice. Meteorites that fall across a wide region of Antarctica all end up here, brought together by the Mawson, Mackay and other glaciers. Here, the brutal winds uncover the gathered gaggle of meteorites, freeing them from the ice to be plucked up by passing meteorite experts. One such meteorite was cataloged as Allan Hills 84001 (ALH84001). Discovered in December of 1984, the meteor was destined to make a splash in the halls of astrobiology.

The meteorite's story began in the formative years of the Solar System. The stone's age measures at an astounding 4.091 billion years, making ALH84001 one of the most ancient pieces of Solar System formation. Researchers established its Martian heritage by comparing isotopes and various elements in the rock (for example, the ratio of iron to manganese) to those found by spacecraft on Mars. In particular, the gas argon has been directly sampled by the Curiosity Mars rover and the Viking landers, and matches the argon found in many meteorites now thought to come from the Red Planet. Of over 60,000 meteorites found on Earth, only ~130 can be confirmed as coming from Mars.[3]

Though ALH84001's tale began on the Red Planet, its adventures took it farther afield. Based on chemical and radiogenic analysis,[4] it appears that the meteorite formed in conditions very different from modern Mars, where temperatures and the environment had running water. Scientists used ratios of carbonate carbon and oxygen isotopes to show that the carbonates in ALH84001 gradually precipitated out of what was likely standing pools of water on or just below the surface of Mars. Then, some 17 million years ago, an asteroid impact sent the rock flying through interplanetary space.

---

[3] The Mars meteorite group is often referred to as the SNC, named after sites in which many were found (Shergottites, Nakhlites and Chassignites).

[4] Including rubidium-strontium, potassium-argon, samarium-neodymium and carbon-14.

It coasted around the Sun for millions of years, finally crashing down onto Antarctica about 13,000 years ago. The environment of Antarctic climate preserved the rock until it was unearthed in 1984, where it sat on a shelf for over a decade (Fig. 6.1).

In 1996, a team of scientists based at Johnson Space Center held a news conference to announce the possible evidence of past life within Allan Hills 84001. Their evidence came on several fronts. First, microscopic structures within the meteorite's interior bear a striking resemblance to microfossils on Earth, particularly those found in samples from the Columbia River Basin in the northwestern United States. Second, the carbon isotopes within those structures appear to be biogenic. Third, the chemical composition of the microbe-like features is a type of organics associated with biology.

Immediately, the scientific rebuttals rained down upon the Johnson Space Center team. Some underlined the fact that the physical structures were substantially smaller than analogous ones found on Earth. Others asserted that the forms were biological, but that they were simply traces of contamination from Earth that seeped in after the meteor made landfall. Still others manufactured similar microstructures in laboratories. But as team leader David McKay pointed out in a later press release, the structures were only one line of evidence, and that any alternative, non-biological explanation had to explain all of the wide-ranging data, including chemical evidence, put forth by his team.

If Allan Hills 84001 does contain traces of ancient Martian life, its implications are enormous. The discovery would show that if life arose on two places in

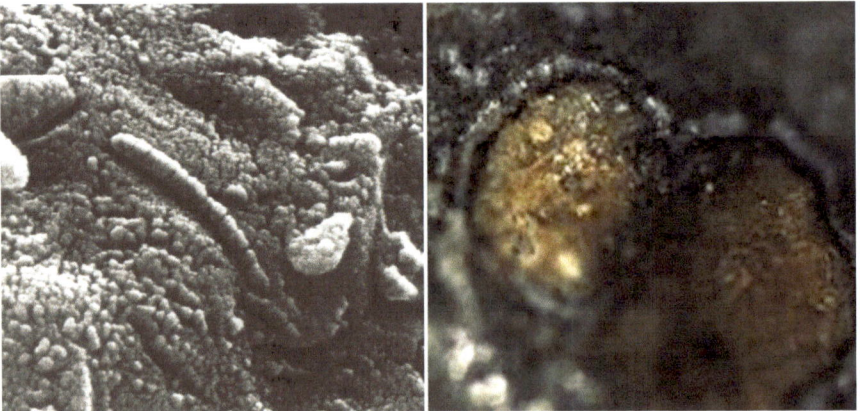

Fig. 6.1 Details of the Allan Hills meteorite. *Left:* Electron microscope images like this one show structures that resemble microfossils on Earth. *Right:* Carbonate globules provide evidence that the Allan Hills meteorite formed in a benign, moist environment. (Both images courtesy of NASA)

our own Solar System, biology in the universe might not be rare. In fact, the stellar neighborhood must be thriving with living systems on exo-Earths that bear only slight resemblance to Earth.

Despite the proclamation's deep implications, public curiosity waned fairly quickly, says SETI Institute's Seth Shostak. "Within a week of this announcement, the general public lost interest in the story. This happened partly because other researchers disputed the claims; many of them doubted that the squiggly forms seen in the meteorite from Mars were ever alive. But there was another reason that this story didn't set the world on fire: microbes are, to most folks, not particularly captivating." When it comes to life beyond Earth, the intelligent brand is far more popular than the microbial type. Still, you cannot have one without the other.

Whether Allan Hills represents the first definitive evidence of off-Earth life, Mars still holds the imagination of the public and the astrobiologists. Peroxides have been found on the Martian surface, and two researchers[5] have proposed that microbes might utilize a mix of hydrogen peroxide and water to live at the chilly temperatures of the Martian landscape. This mix remains liquid down to $-56\ °C$, and tends to attract water directly from the vapor in the atmosphere. They point out that since hydrogen peroxide becomes unstable when heated, the life forms utilizing it would have been destroyed in all the 1976 Viking lander life experiments, which were considered inconclusive.

When the Viking landers first arrived on the Martian surface, the first Earth vehicles to make it down safely,[6] the biology experiments they carried went to work. Within a few months, technicians on the ground had word from the Red Planet. Initially, it seemed that the experiments returned positive identification of active biology. But further study showed no organic material in the soil, and chemistry in the Martian dirt was so reactive that it mimicked biological activity.

Viking carried four life experiments. Engineer Gilbert Levin designed the Labeled Release experiment. LR sprayed a sample of Martian soil with seven amino acids, each tagged, or "labeled," with radioactive carbon-14. The concept was that if any microbes in the soil breathed out carbon dioxide after "eating" the broth, carbon-14 would show up in the soil after the broth was boiled off.

---

[5] See Houtkooper's 2006 paper "Retrocausation or extant indefinite reality? Frontiers of Time: Retrocausation—Experiment and Theory" from the Proceedings of the 87th annual meeting of the AAAS Pacific Division.

[6] The Soviets almost made it five years earlier with their *Mars 3*, but the craft ceased transmitting only 20 s after it landing, without returning any useful data.

The second experiment was the brainchild of NASA/Ames' Vance Oyama. The Gas Exchange instrument assumed that Martian microbes were in a dormant state in Mars' desiccated soil. First, pure water moistened the soil, and gases in the container were monitored for changes. Then, a "chicken soup" of 19 amino acids and other nutrients was applied, and the gases watched for further changes that might be triggered by active biology. In this experiment, oxygen flowed from the soil when the sample was exposed to water, even before it came in contact with the nutrient cocktail. When it came in contact with the "soup," the soil appeared to decompose organic compounds.

Biologist Norman Horowitz crafted a third instrument, which he called the Pyrolitic Release experiment. Unlike the LR and GEX, Pyrolitic Release added nothing to the Martian soil, instead immersing it in carbon dioxide laced with carbon-14. Biologists assumed that if microbes in the soil were breathing CO_2, they would ingest the radioactive carbon-14, which would be incorporated into the soil sample.

The fourth life experiment was simply to look for organic material in the soil using the mass spectrometer on board. The results were puzzling. While some of the experiments echoed ones carried out on desert or Antarctic soil samples, results were abrupt, as if the reactions were chemical rather than biological. Most exciting of the biology results was the Labeled Release experiment. LR's control experiments returned negative results, as they should have. But when Viking applied nutrient-laden moisture to the soil, activity spiked. After months of analysis, results of the other experiments seemed to reflect negative results, so LR's data was put aside.

Since the Viking missions, our understanding of Mars has evolved and matured. We now know that water-ice exists just a few cm beneath the surface in polar regions. We have seen extensive evidence of ponded water in the past, and have seen its fingerprint in the chemistry detected by our orbiters, landers and rovers. Evidence even points to active ice in rock glaciers, and flowing, briny water across the surface in several Martian locations. Many scientists feel it is time to revisit the Viking results.

Mathematicians have reanalyzed the Viking Labeled Release and other data. Using a method called cluster analysis, their work showed a clustering of the Viking biology experiments in one group, and the control tests separated out in another. To back up their data, the team then added data from Earth specimens of both biological and sterile samples. The extra results clustered just as predicted, with biological results matching up with the Viking biology positives, and the sterile ones matching with Viking's control runs (Fig. 6.2).

Perhaps the most disappointing of the Vikings results was the lack of organics in the soil. Some materials in the data seemed suspicious, but were

Fig. 6.2 A retractable aluminum arm (*right*) fed soil samples into the various experiments inside of the Viking landers. The white boom at left contained the meteorology suite. The Viking biology experiments returned baffling results to the astrobiologists. (Image courtesy of NASA/JPL)

dismissed as coming from contaminants from Earth's clean rooms where the spacecraft were assembled. But another line of research casts a different light on the old Viking results: the Phoenix and Curiosity landers have discovered a type of salt called perchlorate. Some investigators now claim that Viking did, in fact, find organics, but the discovery was masked by the perchlorate-ridden Mars dirt. To bolster their claim, researchers repeated the Viking LR experiment on soil from Chile's Atacama Desert, where perchlorates are abundant and conditions are among the most Mars-like on Earth. Their results found the same traces of chemistry that earlier analysts proposed were contaminants from Earth.

The new data does not prove the existence of active life on Mars, but it does indicate that declarations of a dead Mars may have been premature. The Viking mission biology suite may have been a better chemistry experiment than biology seeker. Its experiments provide a cautionary tale for future biological investigations. Astrobiologist Chris McKay reflects:

> In retrospect, the way I would characterize the experiments is that they asked the wrong question. The question we asked with Viking was, "Is there anything alive on the surface of Mars?" We looked for growth, metabolism, etc. But when we go to Enceladus, for example, we're asking the question "Is there evidence of life?" It's not the same question. A dead rabbit is not alive, but it's evidence of

life. A single DNA strand is not alive, but it's evidence of life. We're asking a very different question. We're not trying to culture and grow things; that's the way things were done in the 50s and 60s. Biology is much more biochemically oriented. Rather than trying to grow things, we're looking at their molecules.

Looking for Life: The Future Explorers

With new understanding of the nature of life, and armed with a new generation of tools, astrobiologists are ready to try again at Mars. First up on the new agenda of advanced Mars robots will be the European/Russian ExoMars 2016,[7] the Trace Gas Orbiter. ExoMars 2016 is designed to search for biologically related trace gases in the Martian atmosphere, such as methane. It will also study the surface in an attempt to determine just where those gases are coming from. In the following launch opportunity, flight engineers hope to launch another ESA/Roscosmos mission, ExoMars 2020, scheduled to touch down in 2021. ExoMars' drill-carrying rover will drill down some 2 m to reach samples protected from the intense radiation on the Martian surface. The top landing site choice is Oxia Planum, a smooth plain with what mission designers hope is ancient rock. Once on the ground, the rover will put to work a life detection instrument, the Mars Organic Molecule Analyzer. Drilled samples will be subjected to scrutiny by two spectrometers, looking for hints of organic compounds and molecules associated with active biology. Significantly, the probe carries instruments that will look at the chirality, or structural spin, of any organics (see "Biosignatures" below). An even mix of left- and right-handedness will suggest a geological source of the organics, while a preponderance of one or the other will imply a biological derivation. Next up, NASA has slated its fifth Mars rover, similar to the nuclear-powered and highly successful Curiosity. The rover will set down in the summer of 2020. While gathering rocks for a future robotic return mission, the rover will also utilize a life-sleuthing experiment called the Scanning Habitable Environments with Raman and Luminescence for Organics and Chemicals, or SHERLOC. Mounted to the rover's capable arm, SHERLOC will observe samples without actually touching them, avoiding any possible contamination from Earth-sent equipment. The instrument shines lasers on the rocks from a

---

[7] ExoMars 2016 successfully launched aboard a Russian Proton rocket on March 14, 2016. As of this writing, it is healthy and in transit to the Red Planet.

distance of about 2 inches. Reflected light may betray the fingerprints of organic chemistry in the rocks. The most promising will be cached and sealed for a follow-on sample return mission, still under study (Fig. 6.3).

Fig. 6.3 The European Space Agency's ExoMars deploys the Schiaparelli lander in this artist's impression. (Image courtesy of the European Space Agency)

With the discovery of thousands of exoplanets, biologists are considering what biomes might exist across this smorgasbord of possible environments. Conditions on terrestrial-type exoplanets range from benign atmospheres to hellish ovens to the anemic air of Mars-like worlds. Solid surfaces may include molten rock, cooler surfaces covered in watery ponds and seas, desert-like environments or frozen wasteland. Since water worlds make up a large proportion of exoplanets, researchers have naturally turned to the question of life on them. While some orbit within the habitable zone, many do not. Still, the traditional view of havens for life has changed with our knowledge of extremophiles on Earth, and varying life-friendly conditions in the outer planetary moons of our own system. Benthic life may thrive in the depths of ocean planets, or even in the hearts of millions of ice worlds. And out there, inside one of those rocky planets, just might be lurking a Horta or two.

Is Carbon the Only Game in Town?

The chemical foundation of all life on Earth is the carbon atom. Carbon works well in living processes for a number of reasons. Chemically, it bonds easily with other elements such as oxygen, nitrogen, sulfur, iron, magnesium and hydrogen. It plays well with others. It can form up to four single bonds at a time, and it has stable double and triple bonds, something no other element can do. Because of its molecular structure, it is compact, so that enzymes can use it readily for metabolic operations. It is also abundant in the universe. Because of its propensity for easy bonding, carbon assembles into long chains of complex molecules. These chains are thought to be necessary for life functions.

Life on our own planet—the only life we know—is based on a foundation of about 20 amino acids, chain-like assemblages of carbon molecules used in biological processes. But over 500 types of these carbon chains exist in nature. The discovery of life forms based on other amino acids would mark the discovery of life forms that had their genesis in a fundamentally different path from ours. Finding life using the same organic materials but with a different chirality (structural spin) would also be such a discovery. The trick of finding alien biology is to discover divergent chemical patterns from those we see on Earth. Another clue would be a level of complexity in structure or chemistry that differs from the background "noise" of geological organics.

Biologists are hard pressed to find any life-based chemical alternative for carbon. The most often cited material, for complexity and versatility in a biological system, is silicon. Silicon has some properties similar to carbon, and it's a close relative on the Periodic Table of Elements. Like carbon, silicon organizes into chains of molecules large enough to carry out biological processes. But it has its limitations. Silicon's chemistry is not as flexible as carbon's; it cannot bond with as many types of atoms. The way in which silicon forms bonds limits the kinds of shapes that its structures might form. Its molecules are large and bulky compared to carbon, so they do not easily bond in groups common to organic chemistry. Still, it is found within Earth's biological processes. Many of our carbon-based creatures incorporate silicon into skeletal or protective structures. Some biologists assert that the arrangement of silicates in clays performed a crucial role in organizing carbon compounds in the formation of early life on Earth. Additionally, silicon compounds behave differently under conditions alien to those on Earth. At temperatures similar to those found on Saturn's moon Titan, for example, silicon polysilanols, related to sugars, are soluble in liquid nitrogen.

More exotic materials have been discussed in the search for alien life. Some metals combine in ways similar to carbon. Titanium, tungsten, aluminum, magnesium and iron can all form microscopic tube-like structures, spheres and crystalline forms of the type found in diatoms. Metallic life might arise under conditions lethal to carbon-based forms. Even arsenic, deadly to carbon-based life, is incorporated into the biochemical functions of some organisms such as algae and bacteria.

Extreme Biomes

A perfect storm descended upon the astrobiology community in the 1970s. It was a tempest that brought the idea that alien life might not need to live in an Earthlike environment. Early researchers dismissed the smaller moons of the Solar System as dead worlds. Then, the twin Voyager spacecraft carried out the first detailed reconnaissance of the Jupiter system, giving us our first glimpses of a previously underestimated natural force: tidal friction. Many moons once thought to be too small for geologic activity are prodded by these forces, resulting in the kinds of geological wonderlands we see on volcanic Io and oceanic Europa.

Europa is the smallest of the massive Galilean satellites (moons named after their discoverer, Galileo Galilei). The icy ball travels halfway around Jupiter each time that Io completes an orbit, and twice for each of Ganymede's circuits. Europa is said to be "in resonance" with the other moons. This means that in terms of the tug of gravity from its siblings, Europa is caught in a gravitational taffy pull between Io and Ganymede, with Jupiter also pitching in. Tidal heating—generated as a result of the moons' gravitational tugging—heats up Europa's interior, though to a much lesser extent than Io's (Fig. 6.4).

During about the same time that geologists were coming to grips with tidal heating, the second component of the astrobiological storm arrived in the form of discoveries of extremophiles, life in Earth's most extreme environments. Says former SETI director Jill Tarter, "In a 2004 paper, Craig Ventner characterized the 20th century as the century of physics, and the 21st century as the century of biology. I would say 'biology of Earth and beyond.' Everything came together: the discovery of extremophiles, our discoveries of exoplanets, and the computers involved, all are giving us points of view and tools that we never had before to explore this question. Just over one career, extremophiles and exoplanets have been game-changers."

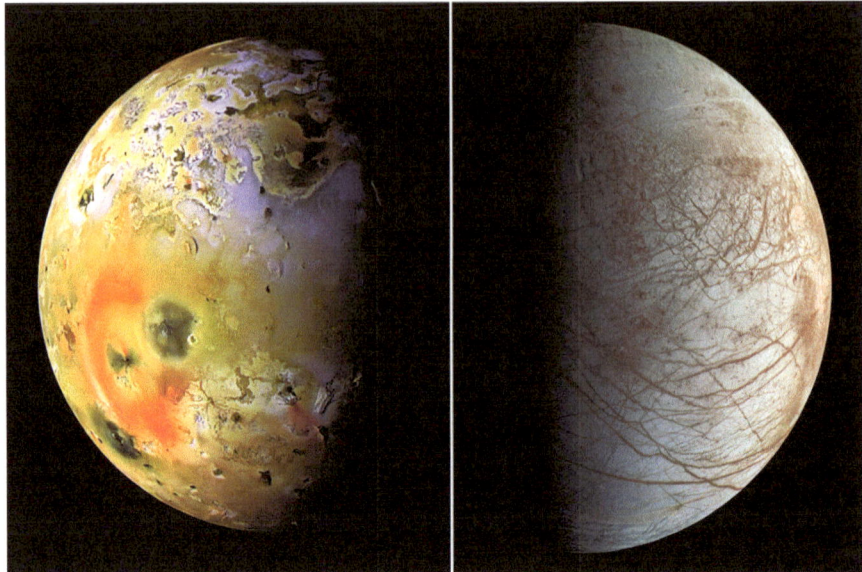

Fig. 6.4 Under stress: two moons under the influence of tidal friction. Io (*left*) is wracked by volcanoes, while Europa's glistening ice surface betrays a subsurface ocean kept warm by tidal heating. (Images courtesy of NASA/JPL)

Under the Surface: Extremophiles Galore

In 1977, explorers discovered seafloor volcanism along the Galapagos Rift zone near the Galapagos archipelago. Explorers had noted hot undersea plumes, but their nature and source had remained a stubborn mystery. On a deep deployment, the robotic submersible *Alvin* revealed spectacular chimneys of sulfur compounds rising from the ocean floor. Researchers began to understand that the number of volcanoes on the ocean floor must outnumber the active ones on the surface.

Undersea hydrothermal vents break through along the mid-ocean ridges where Earth's mantle comes to within a few hundred meters of the surface. Seawater percolates through the crust, heading downward. When it eventually makes contact with the 1200 °C magma, the water heats up to 540 °C. The high water pressure prevents it from boiling. The heated fluid makes its way back up through fractures in the rock, leaching minerals along the way. When it finally flows up into the ocean, it is laced with a complex mineral broth. The rich water streams from these vents, building delicate structures of spires and columns. Some of them tower dozens of feet above the seafloor. The vents

Fig. 6.5 Undersea hydrothermal vents take on fanciful forms as plumes build chimneys. These structures rise from an ocean ridge at the Mariner vent field on the Valu Fa ridge in the southwestern Pacific. (Image courtesy of ALRAB/Aquificales Data Warehouse/NSF)

themselves rise up in stony pinnacles, blobs and delicate curtains. Some chimneys grow 20 feet high in just a year. One chimney, called "Godzilla," had grown as high as a 15-story building when it collapsed, but it is building again even now (Fig. 6.5).

With the detection of the undersea volcanoes came the discovery of entirely new biomes around them. Marine biologists discovered that within that eternal darkness, life thrives in a food chain anchored to the minerals flowing from erupting vents.

The seafloor vents of sites such as the Galapagos and the Juan de Fuca ridge are so deep that the water pressure around them could crush a car like a soda can. But in these high pressures and frigid darkness, where the water is nearly freezing and the liquid squirting from the volcanoes is nearly boiling, a secret garden grows. Giant tube worms nearly 3 m long undulate in the flowing water like prairie grasses. Reddish feather gills stick out of a long white body tube, gathering nutrients from the hot stream that blows from the hydrothermal vents and imbuing the creatures with the appearance of a gigantic lipstick.

A tubeworm has no eyes, mouth or stomach, but simply soaks up the nutrients coming from the mineral-laden stream.

This underwater extremophile menagerie holds even more surprises. Half-meter-long clams bask in the boiling brew, kept company by strange one-eyed shrimp (their one eye is on their back). Blind crabs scamper around in the darkness, munching on the clams. But the hottest inhabitant in these alien biomes is a tiny floating worm. The 12-cm-long Pompeii worm wears a Mohawk of frilly bacteria on its back, and lives in water as hot as 80 °C, the hottest temperatures yet recorded for in-situ creatures.

Two prime candidate sites for current or past life in our Solar System are Jupiter's Europa and Saturn's Enceladus. They exist in the chilled environs of our outer planetary system, while another Saturnian moon, Titan, offers a glimpse of a prebiotic world, an environment whose biological processes were stillborn, frozen before they could become fully living. Astrobiologists put the likelihood of life on a fourth candidate, the planet Mars, somewhere in between.

The rich biomes surrounding undersea volcanoes exist independently of the Sun's energy. Could Jupiter's Europa or other distant moons harbor life forms akin to these?

Europa

As Jupiter's fourth-largest moon, Europa sparkles with a brilliant white surface made of glistening water-ice, inscribed by fractures and faults. Its pristine surface points to a very young age, a scant 50 million years by some estimates. Linear and arcuate stripes paint the surface with telltale signs of powerful tectonic forces. Razor-straight lines roll across the frozen landscape, bracketed by long ridges rising 100 m into the black void. Europa's remarkable ridged landscapes have fractured into vast sections that have shifted and rotated before freezing solid again. Still other areas, called chaos regions, appear to have collapsed into a sea-like slurry, freezing into place after fracturing into puzzle pieces, just waiting for scientists to reassemble them. Data from long-term studies by the Galileo Jupiter orbiter (in orbit from 1995–2003), along with new computer models, point to an ocean 100 km deep. Europa is far more a "water world" than any Hollywood movie.

Aside from visual clues to a subsurface ocean, Europa generates a magnetic field consistent with liquid saltwater. The magnetometer aboard the Galileo spacecraft sensed a change in magnetic fields coming from Europa. Experts in fields and particles recognized the hauntingly familiar electrical currents. They mirrored those generated by Earth's oceans, just the kind that would be produced by electrically conducting liquid within Europa's upper ice region. Unlike the electrical currents pouring from Earth's core, Europa's field is induced—it is created as Europa passes through Jupiter's powerful magnetosphere. The presence of the moon's induced magnetic field led researchers to the conclusion that a near-surface conducting layer, such as an ocean with dissolved salts, was the culprit.

The tidal forces that trigger volcanism on Io are at work—to a lesser extent—on Europa as well. Europa's rocky core likely forms volcanic features, venting on the seafloor of the little moon. It is this mix of volcanism and ocean that has astrobiologists interested in Europa. Submerged in eternal darkness, Europa's ocean depths have one advantage for any alien biology: they are sheltered from the fierce radiation of Jupiter. The Jovian magnetosphere acts like a giant particle accelerator. Jupiter's magnetosphere drapes the Europan surface in enough radiation to tear apart any cell walls within moments. Even an unshielded astronaut would receive a fatal dose of radiation in a day.[8] But beneath the ice, radiation levels drop to benign amounts.

The depth of Europa's ocean depends on how thick its crust is. Some studies suggest an extensive, deep ice shell around Europa, essentially solid down to many tens of km. Other models posit an ice crust of less than 15 km. The structure and thickness of the Europan crust is a hotly debated and complex issue. Some researchers assert that impact craters and jumbled "chaotic zones" indicate a thin, 2-km crust at the equator. The crust appears to thicken to the north and south. But a thicker covering might be explained by the phenomenon of diapirs, masses of ice that mimic a solid version of a lava lamp. Diapir movement could also explain the characteristics we see across the face of this frosty moon.

Whichever model turns out to best describe the crust, evidence indicates that Europa has a global ocean, a vast subsurface sea running from pole to pole under the ice. Scientists are able to piece together a picture of Europa's interior by the way Europa's gravity affected the path of the Galileo spacecraft. As Europa's gravity caused the spacecraft to speed up, its radio signal shifted, just as a siren's tone on a passing fire truck shifts. Europa's internal structure caused subtle changes in this "Doppler" shift with each pass, enabling scientists to chart the moon's interior. Doppler data from the closest flybys fits a rocky interior capped by an outer layer of water 100–200 km deep. This depth is twenty times that of the deepest Earth ocean,[9] and akin to estimates

---

[8] Average levels of radiation on Europa are estimated to be 540 REM per day. The fatal dose for a human is about 500 REM.

[9] The Marianas Trench is 10,999 m deep.

for some of the water worlds such as Kepler 62e and f. Data from this and other observations indicates that the ice shell of the Galilean moon is decoupled, moving independently of Europa's interior rock core.

In late 2013, a series of Hubble Space Telescope images apparently revealed water vapor hovering over the southern hemisphere of Europa. This cloud bore some resemblance to the vapor vented by geysers on Saturn's moon Enceladus, and appeared to be issuing from Europa's own geysers. Given the amount of water in the cosmic cloud around Europa, some investigators estimated the moon was erupting a stunning 7 tons of water every second. The great water plume reached an altitude of 200 km. To create such a cloud, geysers must be erupting at an estimated 700 m per second, three times the speed of a passenger airliner. But a mystery remains—the plume seems to have disappeared.

Scientists are awaiting further confirmation, but the purported discovery would demonstrate that liquid water is close to Europa's surface. It may lie in localized ponds within a thick crust, or the vapors may be direct links to the deep ocean below, gurgling away beneath a thin crust. If the sites of eruption can be isolated, they will be prime targets for tomorrow's astrobiologists seeking answers to what lies in those oceans beneath Europa's icy landscape.

Although water-ice is the dominant substance on the surface, several important materials mix in with the ice, and these materials may be leaking from the ocean below. Amorphous, often radial, discolorations along fractures and in hollows may be the fallout from geyser-like activity like that possibly detected by Hubble. Many of these "painted" features blemish Europa's face. Most are associated with fractures or faults. The rusty stains appear to be endogenic, generated from the inside.

Often, dark material blankets the surrounding terrain, more like particulate matter than a flood of liquid. Just what is this dark material? Instruments aboard the Galileo spacecraft gave researchers enough data to identify several candidate substances, including various salts, with magnesium sulfate (like Epsom salts) or chloride salts as the best spectral match. However, some investigators argue that Europa's spectra are better matched by hydrated sulfuric acid. Though salts are not brown, sulfur is. The inner Jovian environment is bathed in sulfur, thanks to Io's eruptions.

What really gets the astrobiologists' attention, however, is the reddish material, darker than the surrounding landscape. This may be the organic expression of briny subsurface lakes. The briny ice bolsters the idea that dark materials are seeping—or erupting—onto the surface.

Earth biomes at sites such as the Galapagos Rift zone are completely cut off from the Sun-centered world above. Their entire food chain depends on the witch's brew of chemistry emanating from the volcanic vents. Perhaps Europa has developed a similar biological chain. If so, we may find the first alien life in our Solar System on this bizarre ice moon.

Enceladus: Europa on Steroids

Europa made waves in the astrobiology community, but scientists would soon see something even more remarkable—a smaller moon with far more extensive activity. Enceladus is a remarkable world. A tangle of twisted ridges abut cracked plains nearly devoid of craters. These plains appear to have been resurfaced in the geologically recent past. Other regions are heavily cratered. All surfaces of Enceladus are among the brightest terrain in the Solar System, intimating that the entire moon is dusted with fresh material.

A scant 504 km across, Enceladus would barely span the country of France. Because of its diminutive size, Enceladus' geologically young surface mystified researchers. Neighboring Saturnian moons were cold, dead places, despite the fact that many were larger.

Circling close in to Saturn, Enceladus races around the golden giant once each 32.9 h. The orbit of Enceladus is out of round, comparable to that of Io, so some researchers assumed that tidal forces might be strong enough to trigger some kind of internal activity. Other geologists suggested that its interior might be heated by a wobbling motion of the satellite caused by the tug of nearby moons.

The small ice ball has a big tale to tell. Enceladus floats within a gossamer donut of fog surrounding Saturn. Saturn's faint E ring actually envelopes the orbits of several of Saturn's moons, well beyond the main rings that made Saturn famous. Scientists knew for some time that a mysterious source was continually replenishing the fog. Observers also discovered that the ring was dominated by very small ice particles. Such tiny flakes should only last for decades to centuries; something had to be recharging the faint torus. Some researchers suspected that the ring's fine particles were somehow related to Enceladus itself. They were right, but no one suspected how extraordinary the relationship was.

In 2004, the Cassini spacecraft settled into orbit around the ringed giant. Immediately, the robot felt the effects of the E-ring: the environs of Saturn are inundated with floating atoms of O_2. A trio of flybys from February to July of 2005 finally confirmed the presence of cryovolcanic (super-chilled) activity. The imaging science team first spotted curtains of fine mist against the dark sky. Magnetometer readings confirmed the discovery, detecting ions streaming from the moon's rarified atmosphere. Temperatures in the plumes measured as high as −136 °F (some 200° higher than the surrounding environment). This temperature is consistent with a mix of water and ammonia. Cassini's Ion and Neutral Mass Spectrometer detected ammonia in the plumes.

Flight engineers modified Cassini's orbit to pass within 168 km of the surface. Team members wanted detailed data on the magnetic fields and a shot at more detailed geyser images, but they got more than they bargained for. Cassini sailed directly through an extended plume of material. The spacecraft detected 90 % water vapor, with traces of carbon dioxide, methane, acetylene, propane, and possibly carbon monoxide, molecular nitrogen, and soupçons of quite intricate carbon-rich molecules. In addition to all that chemical excitement, the ice particles in the plumes contain sodium chloride (ordinary table salt) and other salts. Salty ice is difficult to make unless it is flash-frozen from saltwater. The plumes appear to be bringing up salty ice grains from the interior. The frozen spray demonstrates that salt water exists not far below the surface of Enceladus at present, occasionally rocketing into the airless sky of the glittering white moon. Something complicated is going on in the chemistry beneath that ice.

The plumes erupt from a series of canyons and ridges in the southern hemisphere. A sunken plain encircles a quartet of darkened valleys called "tiger stripes." These 100-m-high ridges bracket rifts that plunge some 500 m deep. Each is about 2 km across and up to 130 km in length. Dark material extends several km to each side and is apparently erupting or seeping from the rifts.

Scientists can tell what a surface is like by measuring an effect called thermal inertia, the surface's resistance to change in temperature. Cassini demonstrated that the thermal inertia of the surface in Enceladus' southern regions is 100 times smaller than that of solid water-ice, inferring that the landscape is "fluffy," covered in fresh ice or snow. The interior chemistry of the tiger stripes fascinates geologists and astrobiologists alike. The majority of Enceladus' face is composed of almost pure water-ice. But in those chasms, as in the plumes, Cassini's instruments detected organics and CO_2. Some researchers contend that this makes Enceladus the best target in the Solar System in the search for prebiotic conditions or even active life. "Whatever it has in its subsurface ocean is there for the asking," says Carolyn Porco, imaging team leader on Cassini. "We strongly believe the solids are flash-frozen droplets of salty-liquid water that have organics in them, and who knows? They may even have microbes in them. Organics are surely along the tiger stripe fractures."

Observers estimate that geysers on Enceladus spew 150 kg of water into space every second. While the liquid may not escape at a steady pace, the

Fig. 6.6 Canyons rending the tortured southern hemisphere of Enceladus are the source of its illustrious geysers. (Image courtesy of NASA/JPL/SSI)

extent of water in Saturn's environment shows that current levels of geyser eruptions have lasted for at least 15 years.

Another Enceladus enigma is why the geysers are centered on the southern pole. One possible explanation is that the heating started in another location, and that material was transported away from the heated area either by melting of the ice or by loss through eruptions. When a spinning object loses mass in one area, its rotation becomes unstable until the spin axis realigns itself into balance. The active region on Enceladus may have begun somewhere else, and then realigned itself with the pole, becoming stable again. In fact, recent data confirms that Enceladus, like Europa, has a decoupled, free-floating ice crust (Fig. 6.6).

Some researchers prefer Europa over Enceladus. They assert that Europa's ocean is so deep that it must be a permanent feature. The ocean of Enceladus, they contend, is shallow enough that it may be a short-lived phenomenon that comes and goes over short geological time periods. This would shorten the time available for active biology to take hold, they say. But with a subsurface sea not much more shallow than that at Europa, and with its complex organics and low radiation compared to the Galilean satellites, Enceladus is a prime candidate for a home to extraterrestrial life.

Titan: The Alien World Next Door

Saturn's moon Titan is larger than the planet Mercury. Beneath its surly orange fog, ice mountains rise from dark plains. Titan's lowlands are dusted with organic material, the product of interactions between Titan's atmosphere and sunlight in its upper reaches. Methane rains fall down from the sky, carving out canyons and settling into great seas. Those seas mix with the organic material, generating a slurry of carbon-based muck similar to the chemistries of life.

Other exo-Earth biomes might be found with conditions similar to those at Saturn's moon Titan. Despite cryogenic temperatures there (hovering around −150 °C), Titan's environment generates materials that might be incorporated into exotic metabolic pathways, including acetylene, hydrogen and heavier hydrocarbons that drift from the sky. A recent National Academy of Sciences report[10] considered Titan's environment as meeting the requirements for living systems. Organic reactions in hydrocarbons are nearly as flexible as those in water. If so, this expands habitable zones around stars substantially.

Titan's frosty wilderness displays a variety of alien conditions. A deep nitrogen/methane atmosphere blankets its surface. Sunlight interacts with that air, converting methane into rich organic materials that float down to the surface as a fine hydrocarbon haze. The hydrocarbons bank up into dunes in Titan's global winds, draping across a landscape of water frozen to the consistency of granite. Methane rains from the sky, carving river valleys, and pools as great seas. The largest of them, Kraken Mare, rivals the Black Sea in extent.

A curious analogy forms between salt and freshwater on Earth, and methane and ethane on Titan. When methane evaporates, it leaves behind ethane with traces of butane and propane. It's a backyard barbeque on a planetary scale, and within those alien chemistries lie the building blocks of life. Aside from temperature, some biologists believe that conditions on Titan today mirror prebiotic conditions on our own world some 4 billion years ago. Hydrocarbon solvents provide an environment that encourages the synthesis of organic chains such as amino acids. And like the early Earth, before the cyanobacteria oxygen revolution, Titan is devoid of gaseous oxygen. This lack of oxygen and free water also preserves such reactions (Fig. 6.7).

The bizarre, methane-dominated environment of Titan may find a rough analog here at home, on the seafloor in the Gulf of Mexico. A km beneath the surface, under extreme pressure and temperatures near the freezing

---

[10] Baross, J. A., et al. *Limits of Organic Life in Planetary Systems*, National Academies Press, 2007.

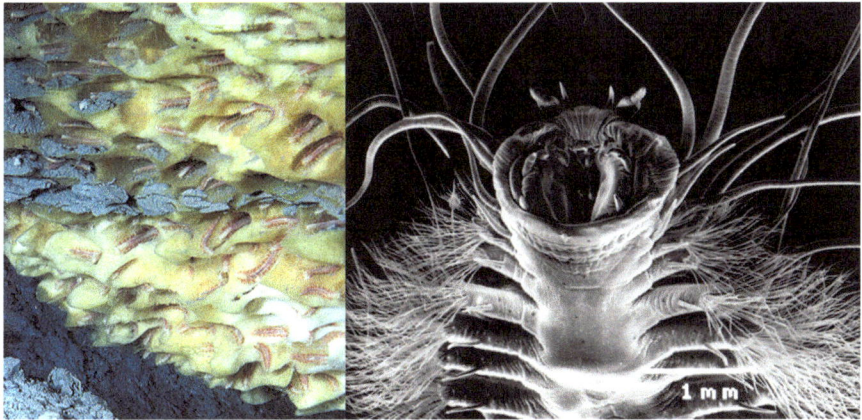

Fig. 6.7 Pink worms form frigid colonies on the yellow surface of methane ice in the Gulf of Mexico. (Image courtesy of NOAA/NASA)

point of water, an entire new ecosystem has been found on ice outcroppings. As methane gas leaches from the rocky flooring of the ocean bottom, it seeps into the frigid ocean water, where it freezes into mounds up to 3 m across. On the honeycombed surfaces of these great yellow mushrooms of ice, explorers have spotted entire colonies of pink worms. Each 2-inch creature, called *Hesiocaeca methanicola*, uses rows of oar-shaped projections to move along the methane ice surface. Biologists suggest that the creatures may be grazing on bacteria that feast on the methane ice itself.

"Titan is wild, and it's also nearby," says NASA's Chris McKay (Fig. 6.8). "We can go and check it out. We may be able to go there and do missions to test our wild ideas. Titan is as alien as you can get in our Solar System and still do something interesting. It would be really cool to find something happening on Titan, because that would tell us that life is not just common, but that it can be bizarre; it's not all cookie-cutter stamped-out-of-water-based life like us, but there are things that are really different."

Only decades ago, Titan would not have been considered a candidate site for life. Nor would the other frozen moons of Jupiter, Saturn, or beyond. But revelations about microbiology, extremophiles, and habitats on distant worlds of our Solar System have led to an expansion of our concept of habitability. SETI astronomer Jill Tarter comments that, "When I was a student, I took any number of classes where the textbook said life can exist between the boiling and freezing points of water in about one bar of pressure, with a neutral PH. Now we're finally giving microbes the respect they deserve…microbes are extraordinarily fit for environments that humans are not suited to at all.

Fig. 6.8 The exotic environment of Titan may spawn pre-biotic chemical reactions. Here, special filters on the Cassini spacecraft's imaging system allow us to see through the hydrocarbon haze to the methane seas and lakes in the moon's northern provinces (*upper left*). (Image courtesy of NASA/JPL/SSI)

There's a lot more potentially habitable real estate out there than we might have imagined. We're finding creatures capable of doing all kinds of ingenious things to make a living."

It was not an easy journey, McKay says. "I think the interesting part to that journey is that we were dragged to it by the data. When we discovered these places and we saw what was there, we said, 'Oh, we didn't think of that.' That's the point here: we're not clever enough to think of these things ourselves. We have to go out and look, and nature says, 'look at this.' We were shown it. Deep-sea vents were another example. We see it; we don't believe it, but then we're forced to accept it because there it is. I think that's a lesson for the exoworlds, too. We should expect to be surprised."

Keeping those expanded possibilities in mind, astrobiologists have turned their attention even more diligently toward Titan. Despite Saturn's distance from the Sun, and despite Titan's heavy fog, light levels on the surface of the planet-sized moon are adequate for photosynthesis. But biochemistry on the surface is limited; life will find few elements accessible on a world made of water-ice. For decades, NASA's strategy for finding life beyond Earth has been to "follow the water." In the case of Earth's biology, that water must be in liquid form (even if it is vapor). In our terrestrial environment, water acts as a universal solvent. It lubricates biological structures, and it opens pathways for biochemical functions. But its role as a universal solvent for living systems relies on its liquidity and on its chemical interaction at Earthly temperatures and pressures. In the case of Titan, the liquid permeating the environment is methane. If biologists are to consider life that uses this as its solvent, all of our assumptions about interactions with the environment must be reassessed in a different light. Recent studies indicate that organic reactions in hydrocarbon liquids carry on with as much gusto as they do in liquid water.[11] Perhaps the hydrocarbons so ubiquitous on Titan might be a biological analog to water in Earth's biome (Fig. 6.9).

Fig. 6.9 Strange biochemistry. What "diverse and wonderful forms" might life in a methane pond exhibit? (Art © Michael Carroll)

---

[11] Bains, W. "Many Chemistries Could Be Used to Build Living Systems." *Astrobiology* 2004, 4, pp. 137–167.

As methane reacts to sunlight in Titan's upper atmosphere, photochemicals combine to create organic molecules. A carbon haze drifts downward, eventually blanketing the water-ice surface and mixing with the methane lakes, resulting in even more complex materials. Titan's lakes and seas become rich brews of the stuff of life, complex organic chains floating in methane baths. If carbon-based life is what you're after, Titan is the place to look.

Although the mineral-rich rocky layers of the moon are cut off from the surface by the ice crust, elements such as iron, copper, nickel, sulfur, calcium and sodium (salts) are constantly being deposited by comets and meteors. But life in Titan's alien environment may not even need the kinds of elements useful to Earth's biology. Because of the cryogenic temperatures on the misty moon, water molecules might be used to fill the role that metals play for enzymes within Earthly biology. Water molecules could act as a catalyst for hydrogen-bonded structures in the same way that metals do for redox reactions (oxidation-reduction reactions critical to photosynthesis and respiration) in terrestrial biochemistry.

Terrestrial biology builds structures using proteins. As they interact with water, individual proteins fold into specific shapes necessary for life functions. In methane ponds and seas, no such proteins can exist, but substitutes may include hydrocarbon chains, aromatic ring structures, and carbon nanostructures such as graphene. And with life thriving in a methane sea, the environment will change. On Earth, increased levels of oxygen, carbon dioxide, methane and nitrogen in the air are the direct result of biology. On Titan, the most promising gases for consumption by critters are hydrogen, acetylene and ethane, all of which would affect the balance of gases in Titan's lower atmosphere. A recent NASA Ames research paper concluded, "The simple low temperature life forms and communities envisaged would have very low energy demands and would grow slowly. Life on Titan may be not much more than [chemically simple] reactions encased in azotomes. However, if it had genetics…what a wonderful life it would be: a second genesis different enough from Earth life to suggest that our universe is full of diverse and wondrous life forms."[12]

Titan serves as a sort of bench test for astrobiologists, offering an alien environment rich in organic chemistry. With our understanding of possible life forms in habitats quite alien from our own, many astrobiologists assert that we should expand our concept of habitable zones still further. The classic yardstick for a habitable zone, a region that allows for liquid surface water,

---

[12] McKay, C. P. "Titan as the Abode of Life." *Life,* 2016, pp. 6, 8.

is too narrow, they say. As an example, they point out that liquid water can exist on worlds far too hot for life as we know it, where water hovers at the boiling point.

Habitable Sub-Zones

Permanent habitable zones (regions that remain habitable over the entire course of a star's life) can be broken down into subsets. For example, an "animal habitable zone" might be considered a region in which an Earthlike world can host liquid water, but where temperatures do not exceed 50 °C. For an Earth biome, this temperature seems to be the upper limit survivable for animal life. An even narrower habitable zone might be one that allows for modern humans. This confined field would include only planets that can sustain the cultivation of vegetation to feed billions of *Homo sapiens*. But on the other end of the scale, habitable zones might be far more forgiving.

Researchers Don Brownlee and Peter Ward[13] refer to an expanded HZ as the microbial habitable zone. For microbes, the habitable zone is vast, close in enough to benefit from the energy of the central star, and near enough to benefit from materials with which to build life (Fig. 6.10). The authors explain, "It is nearly the entire Solar System, and it extends temporally from soon after formation of the planets until the present day."

The ice moons of our outer Solar System demonstrate that planetary formation is varied and exotic. The small Uranian satellite Miranda looks as if someone took a planet-sized eggbeater to it. Next door, Ariel—still small as moons go—shows evidence of volcanic eruption and flow. Both Ganymede and Titan add their spheres to the growing number of worlds with suspected or confirmed subsurface water oceans. Taking into account these unusual sites as possible biomes, the habitable zones of other star systems may be vast indeed.

Deep Sea Astrobiology Beyond Europa and Enceladus

Since water worlds make up a large proportion of exoplanets, biologists have naturally turned to the question of life on them. While some water worlds orbit within the habitable zone, many do not. But considering an expanded

---

[13] For more on this, see their excellent book *Rare Earth: Why Complex Life Is Uncommon in the Universe*, Copernicus Books, 2000.

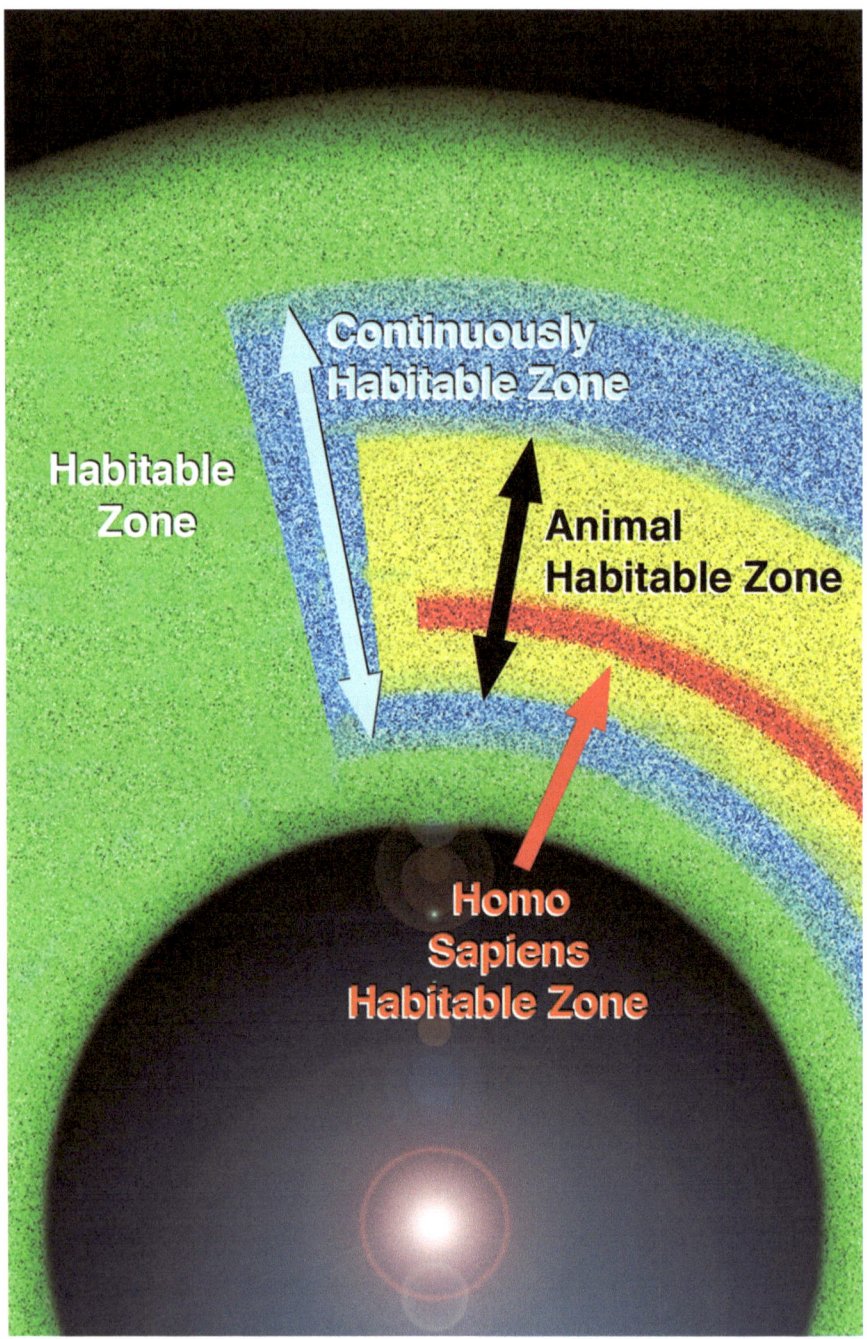

Fig. 6.10 The habitable zone (*green*) can be further divided into (1) the continuously habitable zone (*blue*), (2) the animal habitable zone (*yellow*), and (3) the *Homo sapiens* habitable zone (*red*). The scale of each zone is hypothetical. (Diagram © Michael Carroll.)

microbial habitable zone, our constricted view of life's abodes has expanded with our knowledge of extremophiles on Earth and varying conditions in the outer planets and moons of our own system.

Planets such as Kepler 62e underline the fact that water worlds come in many flavors. Just because a planet has water does not guarantee a comfortable biome. Large super-Earths may have dense atmospheres that preclude habitable surfaces, whether those surfaces are made of land or water. Some oceanic planets orbit on the inner edge of the habitable zone, so temperatures would be high. However, recent research[14] shows that global seas could moderate temperatures as long as the ocean was deep enough. Even in the most extreme case, one which has a planet with an axis tipped on its side like Uranus, habitability is possible, just. In this case, oceans of 50 m or deeper would keep an Earthlike atmosphere warm enough to host life even at the poles. Temperatures there, even with the planet's high obliquity, would remain above 10 °C. But if the ocean shrinks to 20-m depth, a shift occurs. Models indicate that if global waters develop even a thin shell of ice, the entire ocean system could freeze solid in a marine version of a reverse runaway greenhouse effect. One other effect not covered in the study is that of tidal locking. When an ocean world is close enough to its star to be tidally locked, keeping one face toward the star, what will this do to ocean currents and temperatures? Further studies are under way.

A world awash with crashing waves may be a fine candidate for life, if those waves interact with a shoreline. This supposition comes not from the field of astrobiology but paleontology. A study published in the January 2015 issue of *Geology* outlines discoveries of gas bubbles trapped in Earth's sediments representing shorelines some 3.2 billion years ago. At that time in Earth's development, the atmosphere had no free oxygen, so it lacked a radiation-protecting ozone layer. But in this deadly environment, life was able to hang on. Microbes mixed their single cells with grains of sand, forming "biofilms," carpets of gooey biological material.

Some amazing fossils were found in 3.22-billion-year-old sandstone from South Africa. Microfossils from that epoch insinuate that colonies of microbes may have inhabited trapped gas bubbles, where they could endure Earth's austere primordial conditions. Long microscopic channels in crystalline quartz betray the colophon of microbial life that cloistered within the interior of the gas bubbles. The ancient life forms also left a chemical calling card—a ratio of light and heavy types of carbon that is usually associated with living organisms. The creatures left faint impressions of cell chains similar to modern bacteria. If microbes could survive early Earth's harsh conditions hunkered down in

---

[14] "Climate at High Obliquity" by David Ferreira, John Marshall, Paul A. O'Gorman and Sara Seager. *Icarus*, 243 (2014) pp. 236–248.

bubbles, it is possible that life survives on the bleak surfaces of high-radiation planets near M stars or even more exotic suns such as white dwarfs or neutron stars.

Life with a Few Suns

When Luke Skywalker stood at his uncle's doorstep on the planet Tatooine, gazing into the sunset, deep in thought, he looked upon a horizon crowned by a setting sun much like ours. A second, much redder sun followed it down the sky. Luke seemed to be used to double sunsets. He turned around and went to bed with no comment.

Luke's sky was not pure George Lucas fantasy. The majority of stars in the galaxy reside in multiple-star systems. In this way, our Sun is in the minority. Triple-star systems such as Alpha Centauri are the most widespread. Double-star systems, called binary stars, are also common. The brightest star in Earth's sky, Sirius, is a binary star, though its secondary star—the white dwarf Sirius B—is too small and dim to see with the naked eye.

Some stars orbit so closely to each other that they exchange material, creating waves and flares of superheated material and radiation in their surroundings. These systems are probably not good candidates for Earthlike exoplanets. Other stars orbit each other in long ellipses, and would subject any planet to wildly changing conditions. The amount of energy a planet receives from its host star is called insolation. Consistent insolation is widely thought to be critical to life. Biologists look for stable conditions, but these are hard to find in planetary systems with multiple stars. If the stars orbit closely to each other, planetary orbits will be unstable. Stellar gravitational effects will either pull planets in to their destruction or toss planets out of the system completely.

Planets that orbit in the same plane as the multiple stars have another problem: they will suffer varying insolation as one or more stars eclipse each other. Additionally, the brightening of a star as it evolves is compounded by multiple suns. With two or more stars changing in brightness, habitable zones would shift at alarming rates. Still, some stars orbit each other at a distance great enough that stable planetary systems can form, and the effects of secondary star insolation are reduced to reasonable levels.

The skies of distant Earths with multiple suns would see some spectacular phenomena. Earth's familiar rainbows encircle one point in the sky opposite the Sun. But on a world with several suns, rainstorms might generate multiple, overlapping rainbows. Depending on the light emanating from each star, the colors might also vary slightly from one rainbow to another. Days and nights could overlap in strange ways, as suns meet and separate overhead. Seasonal cycles could become quite complex, and if biology has taken hold, vegetation might flower and peak at irregular times. Migratory creatures might make several trips to seasonal nesting or feeding grounds. Whether anyone is there to watch them is an open question.

In the Beginning, Life

Despite our greatest strides in the biological sciences, one of the greatest mysteries remains—how did it all get started? Biologists tend to congregate into two camps when it comes to the origin of life. The first says that life will emerge in any realm possessing Earthlike conditions. Toss in the right amounts of solar energy, radiation, organic soup and a few other minerals, and voila! You've got life. The debate is more complex, of course. For example, did life get going in a comfortably warm primordial stew ponded on the surface of a young world? Or were life's chemistries forged in the cold, drifting clouds of interstellar space, bathed in radiation from passing stars?

The second camp posits that life's rise is incredibly rare, but once it gets going, it disperses throughout the universe, whether in spaceships or aboard meteorites such as ALH84001. The cosmic delivery of life to a planet from an outside source is known as panspermia, a concept first offered by British astronomer Fred Hoyle.

Perhaps the instigation of life can only occur within a narrow set of circumstances constrained by chemistry in the environment, star age, gravity, a planetary magnetosphere, the right geological processes (such as volcanism, plate tectonics and erosion), capped off by just the right mix of gas and minerals. But once that rare life takes hold, it can serve as the seeds of life throughout its stellar neighborhood. Hoyle had good reason to suppose that organisms or their precursors could be transported through the void of interplanetary or interstellar, space. Biologists have recorded cases where bacteria in permafrost, thousands of years old, were revived when warmed. Other bacterial spores have been resurrected after lying dormant for long periods of desiccation in dry environments. The bacterium *Streptococcus mitis* may have even survived for three years in a vacuum on the lunar surface.[15] Isn't it possible, then, that viable organic material could make the journey through interstellar space intact?

---

[15] The robotic *Surveyor 3* landed on the Moon on April 20, 1967. Three years later, the crew of *Apollo 12* demonstrated superb navigation skills when they visited the landing site and brought back pieces of the craft. Within *Surveyor's* camera housing, researchers found traces of *Streptococcus mitis*. The microbes became viable again when exposed to the atmosphere. However, some later analysts proposed that the bacteria constituted contamination due to poor clean-room procedures after the mission. However, advocates of the lunar-survivor scenario reply that fresh terrestrial contamination would have left a variety of microbes, rather than the smattering of the single species found.

Both of these camps rely on a planetary surface as the progenitor source of biological material. Astrobiologist Caleb Scharf suggests another possibility. Perhaps, he proposes, life might arise where the environment around a star and within its protoplanets leads to a wide variety of chemical and physical conditions. Prebiotic conditions, Scharf contends, may be an integral part of the natural cycling of materials from nebulae to star formation to planetary birth. In his book *Extrasolar Planets and Astrobiology*, Scharf puts it this way: "It can be suggested that the conditions in a protoplanetary system are such that extensive organic chemistry must take place." In other words, infant solar systems become factories that are naturally predisposed to pumping out biological material. In early solar system formation, billions of planetesimals drift through the cloud of dust and debris surrounding the infant star. Some are icy, like comets, while others are rocky or metallic. As a side effect of stellar formation, radioactive material is abundant; the larger objects contain significant amounts of radiogenic material, heating them from within. The gravity from these floating icebergs and mountains attracts expanding conglomerations of material. As the cosmic bodies grow, some differentiate, with heavier material sinking to the core. But the process of differentiation is sometimes interrupted by the impact of another large body, starting the mineral mixing all over again. This processing plays out again and again, morphing the same kind of organic material that today exists in many comets and asteroids. Even water appears to be ubiquitous in the interplanetary medium. The space surrounding a new star—far from being a desolate void—may be an organic chemical paradise ready to bloom.

Scharf suggests that during the formation of any planetary system, life chemistry has a window of opportunity, a specific period during which the forming planets can host the interplanetary pre-biotic material surrounding them, incorporating it into their surfaces and nurturing it to the point where biology can arise. Although not solving the origin of life issue, the scenario opens up a much wider range of possible sources for biological material than planetary surfaces.

Still, the biogenesis, or beginning of life, is elusive. In 1952 and 1953, Stanley Miller and Harold Urey attempted to find a solution in a bottle. They pressurized a glass jar with an atmosphere thought—at the time—to mimic Earth's early atmosphere, one with methane, hydrogen, ammonia, water, and no oxygen. When the experimenters subjected the primordial air to electrical charges, tholins precipitated out. Tholins are organic chemicals (Fig. 6.11). The experiment resulted in 20 amino acids, the building blocks of life. But did the experiment really shed light on primordial terrestrial biogenesis?

Fig. 6.11 The environment of Saturn's moon Titan naturally creates tholins. Tholins may make up most of the dark material seen at the landing site of ESA's Huygens probe. (Image courtesy of NASA/ESA)

Life appears to have gotten started very early in the history of our world, within a few hundred million years of the time when the planet's surface was molten, says astrobiologist Chris McKay. "It happened quickly on Earth, but we don't have a clue of how you get it going. The paradigm of the Miller-Urey experiment (where you make amino acids) made everybody think that was the solution to the problem. But it's not. It's the wrong problem. The Miller-Urey experiment is how you get the hardware; where you get the bricks or, in computer terms, where you get the silicon. The real problem is, *how do you write the program?*"

The Miller-Urey experiment demonstrated that getting life's building blocks is easy. If we divide life into hardware and software, the hardware equates to amino acids, proteins, nucleic acids and other biochemicals. Miller-Urey showed that the hardware is relatively simple. In fact, researchers have detected amino acids lying around in many environments. But the second part of life, the software portion, is not "lying around," McKay says. "Somebody's got to come up with the genetic code and program it in. Otherwise you're not Darwinian. At the time Miller did his experiment in the 1950s, we hadn't invented computers, so the notion of software wasn't in anybody's thoughts. Hardware was all they could think about. So they said, 'Aha! We have the hardware problem solved, so problem solved! The origin of life is within our grasp.'"

What those researchers didn't realize was that the aspect of biogenesis that they had explained was the physical part. The more complex and difficult part had to wait for conceptual developments from computer science, and the understanding of the difference between software and hardware. Biologists, says McKay, had to realize that life, like a computer, has both. "Just assembling the hardware doesn't make your computer do anything interesting. You've got to have the software, too. How do you do self-programming? Miller showed self-assembling hardware. Maybe there's a similar brilliant experiment that will show us self-assembling software, where you somehow put things together and hit a spark and it starts writing code, just like it made amino acids. It sounds ridiculous, but I bet before Miller's experiment it would have been just as ridiculous to say you put stuff in a jar and hit a spark and you make the elements of life." Biologists today seek a software analog of Miller's hardware experiment. The software solution has been elusive. The reason may be, in part, because our conceptualization of the problem arose only recently. It has come about with our awareness of computer software. The computer has framed our ability to even ask the question of where this software comes from.

The origin of life remains a mystery. Whether it happens easily or rarely is unknown. Astrobiologists know that the biggest challenge to our understanding is how to generate that coding of life, the biological programming that leads to the first spark within a living organism. That breakthrough is still ahead of us.

Life's 'Tells' in the Poker Game of Astrobiology

In the game of poker, a professional player learns to mask any emotion so as not to give away a good or bad hand. Amateurs exhibit hints as to what cards they may be holding. These hints are called 'tells.' Any terrestrial world with a vibrant ecosystem will have 'tells,' giving itself away chemically.

Nature is in constant battle to balance things and bring them into equilibrium. But the spectrum of light coming from Earth, as the only example we have, shows increased levels of O_2 in our atmosphere, one fairly reliable 'tell.' O_2 is a gregarious gas. It doesn't hang around for long periods, because it likes to combine with other elements. Left on its own, the O_2 cocooning our world would be soaked up by carbon-silicate rocks, bringing it into balance with its surroundings. Its elevated presence, at least on Earth, is due to the metabolism of life forms.[16] In the same way, if Earth had no methane-producing microbes, atmospheric methane levels would drop as the methane oxidized into the oceans and lakes. Methane also combines with carbon dioxide to become more stable. Although methane can also be generated by volcanism, volcanic methane will ebb and flow globally as the environment soaks up and processes erupted gases, while the action of living systems will cause methane to remain more stably elevated over time.

Earth's gaseous envelope is not its only biological 'tell.' The albedo, or brightness, of our continents and oceans can be drastically affected by the color and pigmentation of living things such as plants and forests (on the land) and algae and diatoms (in the waters). The hydrological cycle itself can even be thrown off by living things. For example, plankton impacts local climate by pumping out the gas dimethyl sulfide. This gas supplies sulfur particles to the atmosphere, condensing water vapor. Large plankton populations can actually generate their own cloud systems.

Would our current remote sensing technologies be able to detect life on Earth from a distance? A series of tests carried out in 1990 provided researchers with some insights. In order to gather momentum to reach its destination, the Jupiter-bound Galileo probe traveled a series of looping orbits around the Sun, picking up boosts from flybys of Venus and Earth. On one such Earth encounter,[17] the spacecraft was commanded to carry out the kind of reconnaissance it would be doing at Jupiter five years later. The spacecraft instruments detected high levels of O_2, difficult to explain by anything other than biology. Methane levels in such an oxygen-rich environment showed that something was actively replacing methane in the environment. Galileo also found nitrous oxide in abundance, far higher than it would be if it were at equilibrium with the atmosphere. The ubiquitous nitrogen-fixing microbes in Earth's soil are responsible for generating this nitrous oxide. Galileo's infrared eyes spotted shifts in the spectrum of land masses indicating something

---

[16] The concept was first put forth in 1965 by NASA chemist James Lovelock.

[17] Galileo flew by Venus once in February of 1990 and by Earth twice in 1990 and 1992. It reached Jupiter on December 8, 1995, becoming the first artificial object to orbit a gas giant.

changing the surface (reddish zones indicated living vegetation). The spacecraft was also awash in radio emissions that were artificial in origin, clinching the case for a planet with life. The discovery of any intelligence is still controversial. Galileo's lessons are that astronomers can use technology available today to detect the trace fingerprints of life beyond Earth.

Life, it seems, has inscribed its signature across the face of our blue globe. This telltale trace is called a biosignature. Our search for extraterrestrial life must, consequently, involve a search for conditions that are out of balance, for gases and light curves that are not in equilibrium. In fact, the presence of life not only keeps things out of balance, but it also responds to variations in its star's energy in ways that are more complex than a sterile, stable world would. A terrestrial example is Earth's carbon dioxide (CO_2) levels. Earth's CO_2 varies consistently with the change of seasons. In the winter, vegetation falls dormant, and atmospheric CO_2 rises. But in summer, when photosynthesis blankets the forests, plains and jungles of our biome, CO_2 levels drop. Long-term changes in CO_2 could come from volcanic or other geological activity, but seasonal changes would raise a red flag to exobiologists.

Seasonal changes in the air might also provide clues as to the surface arrangement of a distant twin Earth. In the case of our own planet, it would make sense that seasonal CO_2 levels would cancel each other out as summer moves from the southern hemisphere to the north and back again. But because of the arrangement of our continents, this is not true. Most of the continental land areas on the planet reside in the northern hemisphere, and most of the real estate is north of the equator, where seasonal change is felt most strongly.

O_2 is considered the "gold standard" for detecting distant biomes, but there are caveats and complexities in looking for life. One study cautions that false positives from high O_2 levels may occur when strong radiation strips O_2 atoms from carbon dioxide. If the radiation is furious enough, the process can outpace the rate at which O_2 recombines with carbon (turning back into CO_2).[18] Even so, oxygen is so reactive that it remains in an atmosphere for shorter duration than other gases.

Some watery exo-Earths in close orbits around red dwarf stars may offer another case of mistaken biological identity. Red dwarfs may bathe nearby planets in radiation that vaporizes surface oceans. In this scenario, the atmosphere would become saturated with water vapor. Hydrogen atoms in the upper atmosphere can be stripped away by radiation, leaving behind a shell of dense O_2 gas, resulting in another biotic mirage. The flip side of the coin

---

[18] See Schweiterman, Edward; Domagal-Goldman, Shaun, et al. American Astronomical Society, *AAS Meeting #225, id.407.08.*

is that a lack of O_2 might present a false negative. Just because a planet has low O_2 levels does not rule out the possibility of an active, global biome. The possibilities of both the false positives and false negatives press exobiologists into considering more exotic living systems.

In the search for life beyond, biologists have two primary modes of investigation. They can either culture samples, thereby growing whatever microbes are extant within the sample, or they can search for biomolecules (various organics, amino acids). But in their search for life on Earths of distant suns, astrobiologists have neither option. They cannot ask the pertinent question "Is it alive?" and they cannot search for molecular evidence of life. Instead, astronomers at a distance must look for the waste products of life.

A universal feature of life is that it changes its environment. It consumes and it pollutes. Humans are particularly good at both, but other organisms are accomplished at the process, too. On our own world, biological consumption and pollution has created a global signature of life that would be visible over interstellar distances. Biological waste products here are oxygen and methane. Oxygen holds the most convincing promise of life's byproduct, McKay explains. "The classic case is oxygen in our atmosphere. That's the one that's got to be produced by biology. Now, there are people who have pointed out that there are other ways to make oxygen, that it's not an absolute marker of biology, and that's right. But it's a pretty good marker."

Life's pollution and consumption might make themselves known, even from light-years distant, in other ways. Life often has associations with color. Photosynthesizing plants are green. Algal-seeded salt ponds are candy-apple red. Like the gases of an atmosphere, life tints the color around it. This "pollution" of the spectrum may be a good signature to search for. "Life is flashy with pigments," McKay observes, "and there may be a tiny chance that you could detect such pigments and colorations over interstellar distances."

A third test, one that is almost within our current capabilities, is to look for chirality. Most life forms known to science have a twist to them. This turning format is asymmetrical in a specific way. If the form is flipped, it does not become a mirror image of itself. The phenomenon can be seen in a person's hands, from which the Greek word comes.

The effect was noticed by Louis Pasteur. In 1849, Pasteur was filtering out tartaric acid from wine lees. He noticed that the acid from the living matter rotated the polarized light passing through it, but tartaric acid that had been synthesized chemically (non-biogenically) had no such spin, despite the fact that its composition and chemical reactions were the same. Pasteur showed that the acid's crystals present in two asymmetric forms that form mirror images, one that polarized its light clockwise and the other counterclockwise.

Pasteur realized that the molecules were asymmetric, resembling each other in the way that left and right hands do. More importantly, he noted that the organic molecules consisted only of one type of polarization. Non-organic material showed no such polarization.

Although the effect of this "handedness" is at the molecular level, it is visible at the level of imaging, McKay explains. "It's so speculative that only one or two papers have been written on it. People have talked about looking at reflected light, looking at its polarization and determining the chirality of the vegetation. Jeez Louise in the trees, can they really do that? Again, it's these clever astronomers stretching out data."

One of those rare research articles comes from the *Journal of Quantitative Sectroscopy and Radiative Transfer*. In it, lead author W. B. Sparks says, "All known living material exhibits a remarkable property, which is homochirality—the complex chiral molecules of biology occur almost exclusively with only a single-handedness …For example, the vast majority of organisms use only left-handed L-amino acids in proteins and right-handed D-sugars in nucleic acids. More generally, a chiral compound contains one or more asymmetric centers and thus can occur in non-superimposable mirror-image forms." The paper's authors propose that this chirality, the property of twisting in mirror image, may be a necessity for self-replication. If it is, it is likely to be a generic property of all biochemical life, "whether similar to terrestrial or not."[19] The authors point out that microbial photosynthetic organisms can produce substantial levels of circular polarization in their spectra, while control minerals do not.

The polarized light of chirality could provide a convincing biosignature over interstellar distances. McKay and other astrobiologists recommend those three fundamental steps in the search for exobiology: look for oxygen, look for a chlorophyll red edge or something spectrally similar, and then look for polarized light from chiral-selective microbes or large-scale vegetation. But how would the process of photosynthesis look beneath the sunlight of an alien star?

Finding Exo-Veggies

If plants and trees on an exoplanet carry on something like photosynthesis, taking in the sunlight of the red dwarf star and converting it to food energy, their pigmentation will be different from the green chlorophyll we see on

---

[19] "Circular polarization in scattered light as a possible biomarker," W. B. Sparks, et al; *Journal of Quantitative Spectroscopy & Radiative Transfer*, 110 (2009), pp. 1771–1779.

Earth. M dwarfs emit less visible light and more near-infrared radiation (redder wavelengths) than our Sun. While terrestrial plants absorb the Sun's spectrum, leaving behind green, plants might appear dark yellow or even black to absorb as much of the M dwarf's visible light as possible. Nancy Kiang, biometeorologist at NASA's Goddard Spaceflight Institute, has modeled the light falling upon the landscapes of Earth-sized worlds orbiting their host stars at distances equivalent to habitable zones. Her models take into consideration factors like the star's brightness and color, and the planet's atmosphere.

Using different sample chemistries for atmospheres, Kiang and her colleagues isolated specific color ranges that would be most efficient for photosynthesis. Each sample planet had different dominant colors for the best photosynthetic operations. Kiang found that vegetation on Earthlike planets orbiting stars somewhat brighter and bluer than the Sun might take on a yellow, orange, or even blue tinge by reflecting an overabundance of more energetic blue light. On the other hand, plants on worlds orbiting stars much fainter and redder than the Sun might look black. Researchers at Goddard and other centers have studied light absorbed and reflected by Earth's organisms. Their research reveals that the light given off by planets circling distant stars may betray the presence of Earthlike or non-green plants, like that distinctive red edge found in chlorophyll's spectrum. With studies like those carried out by Kiang, astronomers will know what to look for in the reflected light of distant Earths.

The Stars Are Out, But Is Anyone Home?

As we think about life among the stars, we are forced to look to the only examples of life and potential life biomes we know of—those within our own Solar System. Here, planets and moons offer a variety of environments, and Earth itself hosts many extreme environments where life has taken hold. A few specific examples of the extremophiles—creatures inhabiting extreme environments—may give us a clearer view of the possibilities among Earths of distant suns.

For decades, biologists thought they understood what kind of water and radiation levels and would preclude living systems. But a bizarre little creature with a porcine face and eight stubby legs changed their estimates. Tardigrades, microscopic creatures also nicknamed "water bears," can survive some of the harshest conditions known. They are resistant to the high radiation that may bathe planets in habitable zones of red dwarfs, zones that must hug their dim parent stars to host liquid water. They are also able to withstand extremely arid environments. When faced with desiccated environmental conditions, tardigrades manufacture certain proteins to stave off dehydration. These protein structures remain fairly limber in the presence of moisture, but when the little critters begin to dry out, the proteins form a protective shell, becoming as hard as glass. The glassy casing dissolves when water returns, freeing the tardigrade from its drought-driven hibernation. Tardigrades have more biological tricks up their sleeves. Other proteins enable them to withstand bitter cold temperatures.

The wide variety of conditions in which water bears can survive provide biologists with more possibilities in their search for the presence of life on Earthlike worlds. So do conditions on the seafloor, at the hearts of nuclear power plants, and in the throats of active volcanoes, where microbes have been found basking quite comfortably in searing heat or fierce radiation.

Still other life forms may inhabit the atmospheres of exo-Earths. In 2009, biologists at the India Space Research Organization published a remarkable discovery: three unknown species of bacteria populating Earth's upper stratosphere. Like the tardigrades, these hearty creatures withstand continual blasts of ultraviolet radiation.

We have seen how the biomes at the throats of undersea volcanoes have revolutionized our view of life's possibilities. Microbes also survive the toxic environments of volcanoes on Earth's surface. One of the most hostile environments in the world lies within ice caves on the flanks of Antarctica's Mt. Erebus volcano. When fissures force hot gas through rock, the ice layers above melt into long caves, some topped with ice chimneys. But with no organics inside, microbes must subsist on minerals such as manganese and iron, getting sustenance directly from the rocks.

Erebus isn't the only microbe-ridden volcano. High in the Atacama Desert, where conditions are more Mars-like than any other site except Antarctica, the region's scant snowfall usually sublimates—turns directly to vapor—before it can melt and moisten the soil. The area's nitrogen is depleted below levels detectable by laboratory instruments. Ultraviolet radiation is high because of the altitude, and temperatures swing from –10 °C at night to 56 °C the next day. In this bitter environment, microscopic organisms somehow hold on to life. They do not appear to carry out photosynthesis, but may get energy from chemical reactions between volcanic fumes and the soil. Gases like carbon monoxide and dimethyl sulfide may play a critical role in the survival of these hardy beasts.

Microbes on Erebus and within the throats of other volcanoes offer yet another option for an alien biome where extraterrestrial life might thrive. Sulfurous springs are known for colorful colonies of algae, and the bacteria *Pyrococcus furiosus* thrives at temperatures of 100 °C—the boiling point of water—in Italy's Vulcano, the mountain for which all erupting mounts are named. Biologists report that microbes colonized freshly cooled lavas on the flanks of Iceland's Eyjafjallajokull in the aftermath of its 2010 eruption.

Within Earth's ocean biomes lie even more alien creatures. Some terrestrial organisms actually see with eyes made of stone. Chitons, armored mollusks that cling to sea rocks, grow eyespots made of aragonite, a crystalline form of calcium carbonate. Biologists have long known that the eyes of the chiton consist of light-sensitive spots, but recent work shows that the structures are far more complex than earlier thought. While a chiton's armor consists of the same stony material as its eyes, the lenses of chiton eyespots incorporate larger grains that let in more light. By affixing chiton lenses to microscopes, researchers found that the creatures could make out rudimentary shapes and movements. These strange rocky beasts anticipated the Horta of science fiction, proving that when it comes to creative life forms, nature will undoubtedly have the last word. And many of those life forms may thrive not on planets but on their moons.

What kinds of moons can we imagine in the planetary systems of other stars? Statistics from our own Solar System suggest that swarms of moons circle the giant exoplanets the size of Neptune or larger. Since host planets that orbit their stars in Earthlike paths within habitable zones, conditions on any water-enshrouded moon would resemble those on Earth. But as we have seen with some of our Solar System's own moons, a location in the star's habitable zone may not be a prerequisite for life. A myriad of factors probably come into play when thinking about life-friendly moons out there. For example, ice moons may well be subject to the same forces of tidal friction that heat Io, Europa and Enceladus. Geological activity within these moons may lead to the kind of subsurface oceans we've seen in our own system. Some moons may be the size of planets, exceeding even the largest of our satellites, Ganymede and Titan. From a purely statistical standpoint, large planets have multiple moons, so there is a greater possibility of finding a lot more habitable moons than there are planets.

We have discovered many giant planets in habitable zones, and a wide range of terrestrials as well. The field seems ripe for the discovery of worlds with thriving biomes beyond our own. The search for life forms on Earths of distant suns is a difficult one. Physical distance complicates our efforts at remote sensing, as does our limited understanding of the very nature of life.

Finding biological signatures among the distant Earths would constitute a paradigm-shifting discovery, changing our views of biology, planetary development, and the frequency of life in the universe. Such a discovery would suddenly constrain another variable in the Drake equation. But as we move farther down Drake's line of variables, we focus ever more closely on the holy grail of extraterrestrial life: intelligent beings in advanced civilizations.

7

Could We Make Contact?

A technician plays miniature golf in the wee hours of the morning, the crackle of radio static in the background. We can tell he's been doing this for a long time; he appears bored. Suddenly, from the blizzard of audio noise comes a ping, a coherent, sequential sound. The technician jolts from his malaise, realizing that he is hearing the first signal from another civilization out there (Fig. 7.1). In the background, we hear—ominously—REM's song "It's the End of the World As We Know It."

The opening scene of Hollywood's 1996 blockbuster *Independence Day* expresses both our excitement and our fears about our first contact with an alien world. Another take on the issue, Stephen Spielberg's *Close Encounters of the Third Kind* (1977), portrays aliens not as sinister predators but as curious and wise explorers of the galaxy. Director/writer Stanley Kubrick (*2001: A Space Odyssey*) told a *Playboy Magazine* interviewer, "Why would a vastly superior race bother to harm or destroy us? If an intelligent ant suddenly traced a message in the sand at my feet reading, 'I am sentient; let's talk things over,' I doubt very much that I would rush to grind him under my heel. Even if [aliens] weren't super intelligent, though, but merely more advanced than mankind, I would tend to lean more toward benevolence, or at least indifference, theory."

In his comic strip *Calvin and Hobbes*, Bill Watterson's character Hobbes expresses the thoughts of some when he says, "I think the surest sign that there is intelligent life out there in the universe is that none of it has tried to contact us."

© Springer International Publishing Switzerland 2017
M. Carroll, *Earths of Distant Suns*,
DOI 10.1007/978-3-319-43964-8_7

Fig. 7.1 Long after the twin Voyager spacecraft fall silent, the voices of their creators will still lie in wait—on the surfaces of golden records—to be heard by any sentient being who might discover our artifacts. (Art © Michael Carroll)

Fermi's Paradox

Watterson's characters raise the same question, in a different format, that Enrico Fermi brought up at lunch all those many decades ago: *Where is everybody?* To Fermi, it was unlikely that ours is the only technological civilization within the hundreds of billions of stars in our galaxy. It seemed to him likely that our civilization is typical, not the most advanced technologically, and not the only one exploring the cosmos. Fermi asserted that interstellar travel is not too difficult for beings only slightly more advanced than we are, so some beings or their robot emissaries must have undertaken such voyages by now.

In the ensuing years, astronomers and SETI experts have put forth a host of explanations. A brief survey of the most popular includes…

Answer #1: A "Special" Rare Earth

One answer to the question "With so many Earthlike worlds, where are all the alien civilizations?" may be that Earth is a special planet, so rare that few, if any, other sentient beings have risen to the point where they can communicate with the outside universe. Although this may seem like a return to the ancient concept of Earth as a special creation, there are other reasons to hold to this view. For example, in their book *Rare Earth: Why Complex Life is Uncommon in the Universe*, Peter Ward and Don Brownlee point out the things that make our planet unique: a large moon, plate tectonics, position in the habitable zone of a stable star, and so on. Their conclusion: while the universe may teem with microbial life, the complex set of circumstances leading to higher life forms on Earth are so unlikely that the generation and survival of advanced civilizations is rare.

Answer #2: Extinction

Over 90 % of all life forms that have lived on Earth are now extinct. Several major extinction events have wreaked havoc on the terrestrial life forms (see Chap. 2). Some may have been due to impacts like the one that decimated ~90 % of all life at the end of the Permian age. Other great die-offs could happen at the hands of abrupt global climate changes, massive volcanic eruptions, cosmic radiation influxes from distant supernovas, or the introduction of disparate species from shifting continents due to plate tectonics. For Earth, mass extinction has been a fact of life. Any Earthlike world is subject to the same dangers, and if chance brings slightly harsher conditions to any of these events, the majority of life could be wiped out. Advanced civilizations might not recover from such events, and would disappear before making contact with others.

Answer #3: Not Interested, Thank You!

The human race has not always been searching for life across the skies. Early peoples speculated upon concepts like the plurality of worlds or life out among the stars, but their main concerns centered upon shelter, the next meal or the next land to explore or conquer. The skies were, from any practical standpoint, off limits. So it may be with other sentient beings throughout space. Intelligent life out there may not have progressed to a point where their technology enables contact. Others may be advanced enough to contact us, but may choose not to out of a simple lack of interest. Just because an advanced civilization knows of the existence of another does not guarantee that they will be inclined to try to get in touch. Critics of this perspective point out that it contradicts the nature of the only sentient race we know: us!

Answer #4: We've Moved On

The characters in the *Star Trek* universe have it made. Rather than a radio message taking 773 years to get from the Memory Alpha base at Rigel back to Star Fleet HQ on Earth, they have invented subspace radio, enabling them to chat across that distance instantly (and not keep their audiences waiting). By warping space, they can travel vast expanses of the cosmos in the blink of an eye. Subspace radio and warp speed are fine tools in an alternate Hollywood universe. Sadly, we have no working knowledge of these kinds of handy technologies. But what if a civilization has, in fact, learned how to communicate and travel across great distances? The technology used would be so alien to us that we might not even recognize it. There may be a host of sentient civilizations on hundreds of Earthlike worlds out there, but they are living in the fast lane technologically, unrecognized by us and no longer communicating or traveling by the inefficient means that we use. Until we figure out subspace radio, we will have nothing to talk about. That, at least, is one explanation for the Fermi paradox.

Answer #5: Sentient Suicide

The answer to the "Where is everybody?" question may also lie in the nature of civilization's evolution. Technological civilizations may typically destroy themselves before or shortly after developing space-age technologies. Advanced races may annihilate themselves through nuclear conflict, biological warfare or global environmental contamination. In their 1966 book *Intelligent Life in the Universe*, astronomers Carl Sagan and Iosif Shklovskii surmised that advanced civilizations would either destroy themselves within a century of acquiring interstellar communication, or would overcome their suicidal tendencies and thrive for periods in excess of a billion years. As Shklovskii later put it, "Profound crises lie in wait for a developing civilization, and any one of them may well prove fatal." Critics declare that the idea of self-destructing sentients is "too restrictive." It assumes some civilizations among all those that arise must make it through the survival bottleneck. The assumption that all civilizations commit species suicide is to assume that self-destruction is a built-in fate common to all living things, as preordained as the supernova death of massive stars.

Answer #6: Natural Quarantine

Advanced alien races may exist out there, but they may be spread too far apart to do anything about it. If civilizations are separated by hundreds or thousands of light-years, conventional two-way communication would be impossible. Even if one discovers the other, either or both alien societies might die out before any kind of exchange could take place. Our SETI searches might be able to reveal their existence, but the distances separating us would preclude standard radio communication or extended travel. One civilization might decide to share its knowledge blindly, broadcasting meaningful information into the cosmos, hoping that those who receive it will benefit (see section "Reaching Out" in this chapter). Some speculate that the galaxy is structured to keep sentient civilizations from contact by simply keeping them at a cosmic arm's length. In this scenario, the speed of light acts as a natural barrier between civilizations that might otherwise contaminate or destroy each other.

Answer #7: Cosmic Menagerie

Wildlife specialists who want to observe creatures in their natural habitats go to great lengths to hide themselves from their subjects of study. Likewise, any alien zoologists wanting to study the quirky *Homo sapiens* of Earth might want to go undetected. Advanced beings might also deem it too dangerous for their own good—or ours—to come into direct contact. Many cases in human history point to the wisdom in this, as the meeting between disparate civilizations has often led to tragedy (the native cultures in the Americas meeting Europeans, the conflict between some Arab states and Israel, or the British Empire's annexation of India). It may be that sentient civilizations have come in contact with Earth but have chosen to hide themselves from us. A related theory postulates that advanced alien groups have isolated Earth, creating a virtual universe around us that appears to be empty of life. Science writer Terence Dickinson observes that, "The most valuable thing we Earthlings have to offer advanced aliens is ourselves in our natural state…I believe our cosmic relatives are aware of us and are observing our progress with interest. Passive observation and nonintervention are the only approaches that would pay reasonable dividends for extraterrestrials."[1]

---

[1] *The Universe and Beyond,* Third Edition by Terence Dickinson, 1999 by Firefly Publishing.

Answer #8: Spectator Sport

Extraterrestrial civilizations may be technically advanced enough to contact Earth but are lurking, listening instead of transmitting. Alien minds may be more interested in monitoring other life forms rather than interacting with them. In fact, Earth's own civilizations may not respond to a detected alien transmission. In Principle 8 of SETI's post detection protocol, the official policy of the Search for Extraterrestrial Intelligence organization is that "No response to a signal or other evidence of extraterrestrial intelligence should be sent until appropriate international consultations have taken place. The procedures for such consultations will be the subject of a separate agreement, declaration, or arrangement." The ability of the United Nations or other world organizations to come to some sort of consensus is anyone's guess. Such a consensus would need to address such questions as "What should our response be?", "How should we construct our message?" and "Who speaks for Earth?" A variation on this theme is that interstellar travel is simply too difficult or expensive. But critics point out that even some of our best visionaries could only predict technological advances by a few centuries. Who knows, they ask, what technologies will propel us (or our minds) into the great starry abyss?

Answer #9: Missed It By That Much

In the grand scheme of the universe, humanity has occupied Earth for the blink of an eye. *Homo sapiens* appear to have been scuttling across Earth on the order of hundreds of thousands of years. Earthlike worlds may have been orbiting red dwarf stars for tens of thousands of times as long. Perhaps they came and went, leaving behind no trace, or traces obliterated by eons of plate tectonics and weathering. American anthropologist Loren Eisely observed, "One wonders if perchance their messages came long ago, hurtling into the swamp muck of the steaming coal forests, the bright projectile clambered over by hissing reptiles, and the delicate instruments running mindlessly down with no report."

Using Waves

If anyone is out there, the first front in looking for sentient life is within the radio spectrum. It's a wide spectrum to choose from, and scientists must decide what part of it is most suitable for communication. The cosmos seethes with radio transmissions, waves of energy coming from many different sources. Our eyes are sensitive to a very small part of this wide band of electromagnetic energy. We see visible light, whose wavelengths are about one millionth the thickness of paper. The waves down in the infrared part of the spectrum, which we feel as heat, stretch out to about the thickness of a business card. We can barely see infrared with our eyes in the glow of a stove burner or glowing metal. Below this are the longer microwaves, which we use for cooking and communication. Still lower—and longer—are the radio waves, some of which are as long as a football field.

Above the spectrum we can see lie the ultraviolet, X-ray and gamma ray waves. Ultraviolet wavelengths close to those light waves we can see cause us sunburn. X-rays are as far across as a molecule of water, small enough to travel through our skin and image our bones. Finally, gamma rays have a wavelength the size of an atom, making them tiny enough to cause damage to our DNA. Gamma rays are generated in the hearts of neutron stars, black holes, the explosions of supernovae or nuclear detonations, lightning, and the gradual decay of radioactive substances in our geology (i.e., uranium). This is the electromagnetic spectrum, and it is the song of the universe.

Within all the bouncing radio waves throughout the cosmos, SETI experts are especially interested in one frequency: a wavelength of 21.2 cm (the shorter end of the radio wave spectrum). This wavelength is seen as a frequency likely to be recognized by other civilizations, because it is the wavelength of emissions from the ubiquitous clouds of hydrogen that drift throughout our galaxy. This frequency is called the "water hole," a radio-quiet region in the spectrum associated with water, a requirement for life as we know it. Other wavelengths have also been proposed.

SETI astronomers propose that alien signals might fall into three categories. Signals used by alien races to communicate with each other across their planet are the first category. These signals are known as "leakage," and include transmissions for local use. Some transmissions might be used for communication between populated centers, while other radio waves might be the result of searches for incoming asteroids, for example.

The second type of signals we might detect would be sent from the home world to a nearby site, such as an orbiting space station, moon or planetary colony, or distant robotic spacecraft. We have already been sending out such directed interplanetary communication since the 1960s. But in the case of advanced civilizations, aliens might establish colonies on the planets and moons of nearby star systems a few light-years away. Communications with these will far more powerful, and should be detectable from our Earth-based antennae.

The third type is messages by design. Beings might be actively attempting to communicate with other civilizations by sending out beacons either to simply call attention to themselves or to communicate information. How would this be done? How would we decode these signals? How would we even recognize them? SETI expert Seth Shostak outlines the simple litmus test for extraterrestrial intelligence. "There's an operational definition of intelligence for SETI, and it's very simple: If you can build a radio transmitter, you qualify. It doesn't make any further assumptions. You don't have to be a critter that likes to make poetry. You don't even have to be a critter. You can be machine intelligence. It doesn't matter. If you can produce a signal, then that's what we're looking for. That's pretty darn simple, and yet in a way it's pretty demanding. It really begs the question: Were the Victorians, for example, intelligent? And by this definition, they weren't."

Jill Tarter carries the idea further. "Basically what we've decided is we're going to look for civilization that does something with some kind of tech that modifies its environment in a way that we can detect over these distances. So we search for radio signals, leakage, optical flashes from lasers that might be for communications purposes or for launching spacecraft. It may be that everybody out there is using something that we don't know about yet. The only strategy is to stay around long enough to get old enough and smart enough so that we learn what we need to do."

What Are the Odds?

As we saw in our introduction, the immensity of the cosmos becomes one of the strongest arguments for life beyond Earth. The Milky Way alone has as many as 200 billion suns, and among those, billions of planets. A %age of those must be Earthlike in mass and location in relation to their host star. But the structure of other solar systems seems quite different from what we are used to, asserts astronomer Mike Brown. "We used to get away with saying 'we just haven't seen anything like us yet,' but because the Kepler data are so good, it's not like we're not seeing systems like ours yet. The fact is that we're seeing a lot of systems, and … it's really clear that most planetary systems are just not like ours."

In fact, astronomers Don Brownlee and Peter Ward have taken a second look at the Drake equation and come up with an even more stringent list of variables. Their updated equation insinuates a less-optimistic outlook than that of Drake. The authors assert that simple life forms such as bacteria are common in the universe, but that complex, multi-cellular life more akin to animals (including even more rare intelligent life) is scarce. Items in bold are additions to the Drake equation. The Brownlee/Ward equation reads like this:

$$N * XfpXfpmXneXngXfiXfcXflXfmXfjXfme = N$$

These are the variables:

$N*$ = *stars in our Milky Way galaxy*
fp = stars with planets
fpm = metal-rich planets
ne = planets within their star's habitable zone
ng = stars in the galactic habitable zone
fi = habitable planets where life arises
fc = planets where life has advanced to complex metazoans
fl = lifetime of a planet with complex metazoans
fm = planets with a large, stabilizing moon
fj = Solar Systems with Jupiter-sized planets
fme = planets with a low number of mass extinction incidents

Like Frank Drake's hypothesis, Ward's and Brownlee's work assumes that simple life may be common. In light of research on Martian meteorites and extremophile microbes, most biologists share some agreement on this point. The sheer numbers of planets in habitable zones make microbial life a statistical likelihood, they say. But Ward and Brownlee suggest that even simple life may not *survive in the long term* as easily as Drake's equation would have us believe, nor does it easily lead to more complex forms. Their new estimate takes into account other important considerations, such as the presence of metals important to biological functions, the length of a stable environment, a stable orbit and/or axial tilt (Earth's is stabilized by our large moon), and extinction events. Being the right distance from the right kind of star is not enough. In their book *Rare Earth*, the authors conclude that although future research will deepen our understanding of life's origins and abundance, "… even at this time, it appears Earth indeed may be extraordinarily rare."

Astrobiologists could make the case that although Earth-mass planets in habitable zones are not as abundant as once thought, the super-Earths hold the potential for life on bigger worlds. But these most common exoplanets, the super-Earths or sub-Neptunes (see Chap. 5), may not be so promising for life after all. An Austrian Academy of Sciences team modeled developing planets in habitable zones of Sun-like stars. They wanted to see how much surrounding gas various planets would hold on to. The team studied sample planets ranging from 1/10th to five times the mass of Earth. Their models suggest that planets the size of Earth or smaller may capture a dense atmosphere of gases such as hydrogen and helium early on, but they will lose the bulky atmosphere later, opening up the possibility of more Earthlike conditions. But in the case of super-Earths, gravity holds on to nearly all of the hydrogen, creating a more Neptune-like than terrestrial world. In the case of super-Earths like Kepler 62e and f, the models showed that the two planets gobbled up 100 to 1000 times the hydrogen contained in Earth's oceans, but lost very little over the long run, leaving them with sub-Neptunian conditions (Fig. 7.2). Super-Earths with low densities are likely sub-Neptunes.

Even if the outlook for life on sub-Neptune worlds is dim, we are finding many planets akin to Earth in size, and some of those lie within the habitable zones of their own suns. It seems likely to many that a %age of those Earthlike

Fig. 7.2 The planets KOI 314c and Kepler 452b are similar in diameter, but Kepler 452b (*right*) is more dense, leading astronomers to the conclusion that Kepler 452b is a more conventional terrestrial super-Earth while KOI 314c is a gaseous sub-Neptune. (Art © Michael Carroll)

worlds must have seen the rise of complex biological forms that would, ultimately, desire to define and explore the universe around them.

Is this really the case, though? Even if we assume that Earthlike environments tend to lead to biogenesis (and that's a big assumption), it may be a naïve assumption to believe that living creatures naturally advance up the scale of intelligence. SETI expert Seth Shostak believes that life out there is ubiquitous. "I continue to bet everybody a cup of Starbucks that we'll find intelligent life within two dozen years. NASA's chief scientist said a couple months ago that she thought we would find life within twenty years. What she's talking about is the kind of life NASA's looking for, which you'll need a microscope to see." In their book *Cosmic Company: The Search for Life in the Universe*, authors Shostak and Alex Barnett argue, "The idea that biological history on Earth is simply the flowering of a 'tree of life,' with humans eventually and inevitably budding at the leafy crown, is popular with the public, but considerably less so with evolutionary biologists. Sixty five million years ago, a rock the size of London slammed into Earth, and its catastrophic after-effects wiped out three-fourths of all species, including the dinosaurs. If this hadn't happened, the rat-like mammals that eventually evolved into *Homo sapiens* wouldn't have inherited the world…" The authors conclude that even if worlds with life are abundant, odds are that their inhabitants may be less like us and more like dinosaurs or cockroaches.

Finding any kind of biology beyond Earth will shed light on the very nature and heritage of life, showing us one of two things: (1) Either life on a nearby world is related to that on Earth, and perhaps brought there by meteor impact and transport, or (2) life got its start independently in another site, demonstrating that life must be ubiquitous in the universe. But biogenesis is a fairly sedate discovery compared to the revelation of an extraterrestrial, technologically savvy civilization.

We are left with the question: Does life inevitably lead to intelligent, sentient beings? If so, does intelligence inexorably lead to advanced technologies that might be attempting to contact us at this very moment?

The first issue is the rise of intelligence. If evolutionary forces are at work, should we assume that given time, evolution will naturally lead to intelligent beings? The fossil record says no. Life's progression throughout the history of Earth has been one of fragmented starts and stops, with various life forms rising to take the place of others that die out gradually or, in the case of the dinosaurs or the Permian biome, vanish abruptly from some extinction event.

Primates and cetaceans arose at various times in various locations, only to die off or develop into something less brilliant. An example is the whale. One measure that biologists use to judge intelligence is the ratio between brain mass and body mass. Marine biologists applied this ratio, called the encephalization quotient (EQ), to 62 species of cetaceans that evolved over the course of 50 million years. Far from a linear improvement, the researchers found that whale intelligence increased 35 million years ago, concurrent with the advancement of echolocation (whale sonar), with another bump 15 million years in the past. But not all EQs went up; intelligence ebbed and flowed across history and species. Intelligence has taken on many forms in the history of our world, but it is clear that time plus biological mutation does not necessarily lead to increased intelligence. Some astrobiologists suggest that ancient worlds such as those orbiting red dwarfs have a better chance of intelligent inhabitants, as their planets have been in existence longer than Earth, but this may not be the case.

The second issue is the rise of technology. Assuming that intelligent beings have arisen among the many billions of Earthlike worlds in our galaxy alone, what are the chances of these beings developing technologies that could communicate with other beings, either by transmissions or travel? This is the sixth variable of the Drake equation, and it is an important one. Some anthropologists assert that, given time, the rise of science and technology is a natural outcome of sentient species. But there are both sociological and biological reasons to argue against this.

The sociological front presents another problem. Judging by brain size, some predatory dinosaurs (such as the dromaeosaurs or other theropods) must have been very clever, perhaps hunting in coordinated packs. The dinosaurs had millions of years to develop technology, plenty of time to figure out radio telescopes and starships. If time is the primary variable, why didn't the dinosaurs travel to the Moon, or leave behind their statues and ruins? (Fig. 7.3)

Scientists estimate that *Homo sapiens* have been on Earth, in our current, modern form, from 100,000 to nearly a million years. Over the last 30,000 years, the level of human intelligence has not changed significantly. We did not gain the ability to communicate or travel through space until the last century, a fraction of the time we've been thinking about campfires and using tools. Science itself has developed through a gradual process over the last 2500 years. Multiple cultures have established various systems of mathematics and astronomy, for example, but only one line of advance led to modern science and technology—the one rooted in ancient Greece.

Casual observation told ancient sky watchers that Earth was the center of the universe, with the Sun and planets circling around it, tracing out their orbits across the sky. But the planets—literally "wandering stars"—did not follow smooth paths, so another explanation was ultimately needed. As early as 260 B.C., Greek scientist Aristarchus put forth another possibility: that Earth goes around the Sun. Most of his work has been lost, but it is clear that

Fig. 7.3 If long life inexorably leads to intelligence and technology, why didn't the dinosaurs go to the Moon? The giant Mesozoic creatures ruled the world for 180 million years, but never developed technology. This painting shows statues of sauropods, long-necked behemoths, rotting away in the sea. In the foreground romps a duck-billed hadrosaur known as a Parasaurolophus. The time period between the sauropods and the later hadrosaurs was as long as the time between the hadrosaurs and the humans, a very long period. Nevertheless, the dinosaurs left no statues or radio antennas behind. (Art © Michael Carroll)

although he was on the right track, his ideas were not popular at the time. Still, his loyalty to marrying scientific theory to physical observation paved the way for the modern scientific process in ways that no other culture did. Nearly two millennia later, Nicholas Copernicus constructed his theory of a Sun-centered Solar System. His work ultimately led to our modern view of the cosmos, the view that the Sun is one of many stars, and that planets— including ones like Earth—may orbit other distant suns. This long chain of events and deductions is firmly rooted within the philosopher/scientists of ancient Greece.

Amoebas to Technology

Did the rise of technology come as a lucky break, a bizarre historical fluke? How rare is this phenomenon? We have only one example, humans, and until we have others we cannot know. It's a question that perplexes many SETI theorists such as Shostak. "Just because you have a species with the cognitive capability that we have, will they in fact ever produce radio transmitters and so forth? That depends on two things: one, that physics is universal. Astronomers proved that a long time ago, so you don't have to worry about that. But the second thing is, will they develop science that leads to the kind of tech necessary to become intelligent. *Homo sapiens* went 200,000 years without doing it. There could be a lot of species that are perfectly happy to subsist and do whatever they do and never get into science. From the standpoint of SETI, the very pragmatic approach is, 'Okay, there may be lots of societies out there like Borneo bushmen that just live, but you're not going to hear from those guys.'"

The real question is, will intelligent species develop science or not? Researchers, including Tarter and Shostak, tend to think the latter, because, as Shostak puts it, "Once you have agriculture, then you have people tied down, with time to develop—if you will—white collar jobs, and white collar jobs inevitably produce people who are interested in understanding things." Many researchers are optimistic that if high-IQ beings appeared in our world, an apparently average terrestrial, then sooner or later, intelligence should arise on other worlds as well.

This brings us back to Fermi question, "Where is everybody?" Now that we are filling in more variables in the Drake equation, the likelihood of intelligent life in the universe might seem less promising. But time may be on our side. Our Sun is a fairly new star. Some of the red dwarfs have been around for twice as long, and we have seen that at least a few of them may well accommodate Earths or super-Earths. These planets have had longer timelines from which intelligent life could arise. This means that intelligent extraterrestrial societies may have been around for far longer than we have, inhabiting such systems as Kepler 444. Technological beings that have been around for millions or billions of years may have developed technologies that would seem, to us, as magic (known as Arthur C. Clarke's third law[2]).

Perhaps it is in the nature of civilization to self-destruct, either by destroying the biome in which it lives or by annihilation via advanced weapons. But surely, if the vast numbers are correct, some must get through, surviving to become spacefaring citizens of the universe.[3]

Suppose that radio transmissions are a common developmental step for a surviving, advanced civilization. If this is the case, we should be able to listen

---

[2] "Hazards of Prophecy: The Failure of Imagination," by Arthur C. Clarke in *Profiles of the Future* (1962).

[3] The scope of this book covers Earthlike worlds, but it is also possible that advanced civilizations will arise on frozen worlds with subsurface oceans, or on other worlds very unlike our own. Whether we could communicate with such different beings is another matter.

in on some of the errant transmissions—called radio "leakage"—of such a group. If they are listening, our first episodes of *I Love Lucy* are just reaching the planetary system of 70 Virginis, which may have an Earthlike moon orbiting a gas giant (any inhabitants of the super-Earth Gliese 422b—or one of its moons—have already been listening in for two decades, giving them insights into the advantages of smoking cigarettes, the latest models of 1950s cars and joys of a good laundry detergent). Such leakage assumes that advanced civilizations continue to use radio transmissions (see Fermi paradox Answer #4).

Why have we not heard any alien reruns yet? One reason may be that most of our signals that fall into the same "leaky" category are aimed at the horizon. Those types of signals are not directed. Consequently, even at the distance of the nearest star, it would be take an exceptionally large antenna to pick up the weak transmission. Another reason for the seeming silence is that "leakage" might not last long in the evolution of sentient life. Astronomer Carl Sagan said, "I think there is just a 100-year spike in radio emissions before a planet becomes radio-quiet again." SETI's Doug Vakoch agrees. "If you look at telecommunications as it is developing here on Earth, we have been noisy in the past. We had a lot of TV and radio going out into space. Now, as we shift to communication by fiber-optic or by telecommunication satellite, there is less of this leakage going off into space." So while Earth's broadcasting began as a brute-force affair, blaring out across the airwaves in all directions, transmissions are becoming more directed, with narrow-beam or fiber optics communication taking the lead. Today, even most television is "broadcast" via cable, something no advanced race could or would want to listen in upon.

Despite the challenges, we might be able to hear leaky communications from such beings, or transmit directed messages ourselves. The latter, in fact, has been tried in several forms.

Reaching Out

One way of making ourselves known to distant civilizations is a technological message in a bottle, a physical artifact containing information about us. In December of 1969, the Arecibo Observatory in Puerto Rico sponsored a meeting of the American Astronomical Society. During a coffee break, Carl Sagan mentioned to Frank Drake that the *Pioneer 10* and *11* spacecraft,[4] scheduled for launch just over 2 years later, was going to grab enough speed from its Jupiter flyby to actually depart the Solar System. This curious fact was first recognized

---

[4] Both got help from gravity boosts from Jupiter.

by journalist Eric Burgess and planetarium expert Richard Hoagland. NASA asked Sagan to construct a message that might be decipherable to any alien civilization that might stumble across the wayward Earth relic in the distant future. Burgess later wrote, "I visualized how *Pioneer 10* escaping from the Solar System would become mankind's first emissary to the stars.... It should carry a message that would tell any finder of the spacecraft a million or even a billion years hence that planet Earth had evolved an intelligent species that could think beyond its own time and beyond its own Solar System."[5]

Drake was intrigued by the idea, and happily accepted Sagan's request for help. The assignment was a formidable one. How would one construct a message that showed the time and location of the little robot's launch, as well as something about the strange creatures who sent it?

At the time, Drake was active in research on pulsars, those spinning stars that beam pulses of energy out like a lighthouse. He realized that these bizarre freaks of nature might be good waypoints from which to locate Earth within the Milky Way. Each star had its own unique frequency, so Drake could mark 12 star "cycles" arranged around a central point, providing a location. Pulsars offered another useful feature: their pulsing frequencies gradually shifted by very small amounts (many of them change by less than a billionth of a second each day). In this way, the pulsar map also became a clock, because any civilization that captured *Pioneer* could see the difference between the pulses on the map and those in their time, enabling them to determine the time elapsed between launch and discovery.

Drake and Sagan wanted to add something more. Sagan added a scale diagram of the Solar System with a line indicating the *Pioneer* trajectory. His wife, Linda, drew the figures of a man and woman standing in front of an outline of the spacecraft for scale. The figures shared characteristics of all races.[6]

The image was ultimately engraved on a $6'' \times 9''$ gold anodized aluminum plate and affixed to the side of both Pioneers, becoming the first attempt at physical contact with an alien civilization.

Two other Earth artifacts still winging their way into the great unknown are the Voyager records. *Voyagers 1* and *2* were far more sophisticated than the earlier Pioneers. Each 1500-lb craft is powered with plutonium for the long, dark voyages. Designed for a decade-long mission, *Voyager 1* encountered Jupiter and Saturn in 1979 and 1981, while *Voyager 2* did the same and followed those encounters with Uranus (1986) and Neptune (1989). The twin explorers had enough weight capacity to carry an extra object. That

---

[5] "By Jupiter, Odysseys to a Giant" by Eric Burgess (Columbia University Press, 1982).
[6] The plaque proved a challenge for newspapers and magazines, many of which airbrushed out several critical anatomical features. Some letters accused NASA of sending "smut" into space at taxpayers' expense!

Fig. 7.4 Earth artifacts for the skies. *Top:* Gold anodized plaques provide clues to our location and the time of the Pioneers' launches. *Bottom left and right:* The cover of the Voyager records contained instructions on how to access its information, and the protected interior record contained photos, messages and songs from the entire globe. (Pioneer plaque image: https://commons.wikimedia.org/wiki/File:Pioneer_plaque.svg; Voyager record images: http://grin.hq.nasa.gov/IMAGES/SMALL/GPN-2000-001978.jpg, http://dayton.hq.nasa.gov/IMAGES/LARGE/GPN-2000-001976.jpg)

object still needed to be quite compact and light, but Frank Drake realized that a souped-up phonograph record could fill the bill, not only carrying an engraved message but also images and sounds with hundreds of times the information capacity as an engraved image (Fig. 7.4).

The Voyager records ultimately ended up with 118 black and white and color images, and a wide selection of natural sounds, including crashing surf, bird and whale songs, thunder, traffic, and vocalizations of many species of animals. Engineers recorded greetings from the president of the United States and the U. N. Secretary General, along with greetings in fifty-four languages. The record's wide variety of music included Bach, Beethoven, Stravinsky and Mozart, Senegalese percussion, a Pygmy girls' initiation song, aborigine songs from Australia, Georgian and Navajo chants, Blind Willie Johnson and Louis Armstrong.

The records were created in a rush, as the planets wait for no one. In the book *Murmurs of Earth*, Frank Drake later lamented, "The short time available to assemble the record caused a regrettable flaw in the end product. The sounds, music, and pictures are all recorded separately…How much better it would be if the human voices were next to the appropriate picture, the sound of a motor next to a picture of a car, or the picture of a violin next to the sound of its music. If ever there are recipients of the Voyager record…[they] may also recognize that the lack of such a mix means one thing: those ancient artists who gave them this record, now dead a billion years, just didn't have time. Too rushed…Perhaps this will be nothing new to Them. Perhaps there will be a motion we wouldn't recognize, to Them a nod, as They realize that a billion years before there had been a civilization little different from Theirs."

By the late 2020s, the twin spacecraft will fall silent, their plutonium-238 power sources weakened to the point where they can no longer keep systems warm enough to operate. But they will continue on, their precious records relatively intact. Experts predict that at the end of one light-year of flight, the outer surface of the records, the one facing exposed space, will be pitted by micrometeorites and other interstellar debris over less than 2 % of their surfaces. But between the stars, the environment is far more empty; at a distance of 5000 light-years away, the records will only have degraded by another 2 %. And the interior side, the one with all the data, should remain essentially unscathed.

Neither spacecraft was targeted toward a specific star, but *Voyager 1* should pass within 1.4 light-years of the M-class red dwarf Gliese 445. Although that star is now nearly 18 light-years away, it is drifting toward the Sun,[7] so that *Voyager 1* will make its closest pass in roughly 40,000 years. At about the same time, *Voyager 2* will come within 1.7 light years of the star Gliese 905 (also called Ross 248).

Speaking about the remote chances of anyone finding these artifacts, Carl Sagan commented that while the record would be played only if there are

---

[7] At a speed of about 119 km/s, or 270,000 mph.

advanced spacefaring civilizations in the cosmos, "…the launching of this bottle into the cosmic ocean says something very hopeful about life on this planet."

One other artifact is leaving our Solar System, and it will carry messages from home. The New Horizons spacecraft coasted by Pluto in July of 2015. At launch, it carried no plaque or record for interplanetary soliloquys. Instead, a message to the stars will be uploaded from Earth. Called the One Earth Message, NASA has offered an opportunity to the people of Earth to contribute written messages, photos and audio files. At the end of the submission period, anyone who wants to will get the chance to vote on the content they think should be uploaded to the probe. As One Earth's promotional materials puts it, "It will be a global project that brings the people of the world together to speak as one. Who will speak for Earth? YOU WILL!" Engineers are still in the process of developing an internet interface for global access. They are also developing a format for the messages that can be decoded by any future beings that might run across New Horizons, wandering silently between the stars (Fig. 7.5).

Fig. 7.5   New Horizons is now coasting through the Kuiper Belt after its successful encounter with Pluto. Soon, messages from the people of Earth will be loaded aboard the craft, ready for its voyage into interstellar space. (Photo composite by the author using images courtesy of NASA/JHUAPL/JPL)

Transmission in a Bottle

The long travel times and random trajectories of our spacecraft make their discoveries unlikely. But if we want to send out a message in a way that is more directed than the message-in-a-bottle technique of the Pioneers and Voyagers, radio is the way. Radio waves travel at the speed of light. If we set out today to send a message to the nearest star system, Alpha Centauri, our transmission would be received a little over four years from now. Their return message, if they were quick about it, would reach us 8½ years from now.

But what would our transmitted message look like? SETI researchers have already determined that we must set out with some assumptions. First, our message assumes that the recipient has a grasp of basic mathematical and geometrical understanding. Designers consider this a strong assumption, as anyone receiving our message must have engineered a technological piece of equipment to collect it in the first place. If our messages are representative in a pictorial or diagrammatic format, the recipient must share at least some similarity with us in terms of visual senses. Within those guiding principles, designers must labor under the notion that other civilizations share an understanding of certain shapes and visual trends. But species with fundamentally different senses of vision may interpret symbols incorrectly. For example, when designers put together the Pioneer plaques, a debate raged about the position of the man's upraised hand. Was it a greeting? A gesture of hostility or aggression? A deployment of some sensory organ?

With SETI's general assumptions in hand, designers would then need to craft a message, turn it into bits of information, and send it out into the wide spectrum of the radio universe around us. Radio waves inundate the cosmos. Most of this radiation washes over our planet in wide bandwidths. A narrow-band signal would get the attention of any radio observer worth his or her salt. If the signal was also intermittent, turning on and off or switching between two adjacent frequencies, we might also suspect that this was an artificial transmission of some kind. Researchers look for repetition and patterns. For example, if the series of radio blips repeated again after 1679 beeps, we might notice that the number was the product of two prime numbers,[8] 73 and 23. We would then assemble the blips in a grid and search for a message, perhaps an image.

Such an image graces the pages of SETI history, but it did not come from another world. It came from this one. In 1974, the world's largest radio antenna in the world took part in one of the few terrestrial attempts to send

---

[8] A prime number can only be divided by 1 and itself. Prime numbers include 1, 2, 3, 5, 7, 11, and 13.

specific messages into the depths of space. Engineers fed the message through the mighty Arecibo radio antenna, a 305-m dish in the mountains of Puerto Rico. The transmission lasted for just three minutes, and is today making its way through the void, destined for the stars in the globular cluster M13.[9]

The Arecibo message was structured exactly like the example above. When its "pixels" are arranged in rows of 23 each, an image is revealed. The top of the graphic shows the numbers 1–10 in the "language" of the graphic. Below this, a series of shapes represent the atomic numbers of hydrogen, carbon, nitrogen, oxygen and phosphorus—the essentials of life on our world. When the following rows are assembled, a clever alien recipient would see formulas for the chemistry making up DNA. Under those rows, a wave of squares unwinds to portray the double coils of DNA, leading to a stick figure of a human. Down the center of the DNA helix is a pair of rows describing the number of nucleotides in DNA. To the right of the human is the number 4 billion—the human population at the time of launch—in units relating to the wavelength of the radio signal. Using the same units, a scale to the left shows the average human height of 5 feet 9½ inches. Stretching across the frame at the feet of the human figure is a schematic of the Solar System, with the large Sun at left and Earth, third planet in, indented toward the figure to show where we live. Finally, the base of the image depicts the Arecibo radio dish itself, with a scale showing its diameter across the bottom (Fig. 7.6).

Radio searches are tricky; they involve the teasing out of subtleties in the spectrum of the signals falling upon us from every corner of the universe. In 1961, the Parkes radio telescope came on line outside of Parkes, Australia. In 1998, its giant dish antenna began to regularly detect radio signals called perytons, rapid bursts of a few milliseconds. Strangely, Parkes was the only receiver to detect such transmissions. Eventually, observers determined that the radio signals emanated from moderately intelligent life inhabiting space quite nearby. Researchers tracked down the source of the transmissions to the Parkes Observatory facility itself. They realized that whenever a person in the local kitchen became impatient and opened the microwave oven before its cycle was completed, a frequency-swept radio pulse burst forth, salting the nearby radio space.

In order to isolate radio dishes from the cacophony of day-to-day human life, researchers must locate radio telescopes in remote locations. Additionally, all cell phones and digital cameras must be shut off. At the Green Bank Observatory in West Virginia, for example, several critical rooms are sur-

---

[9] M13 has roughly 300,000 stars and is 25,100 light-years away. At that distance, the message will arrive in the vicinity in A. D. 27074.

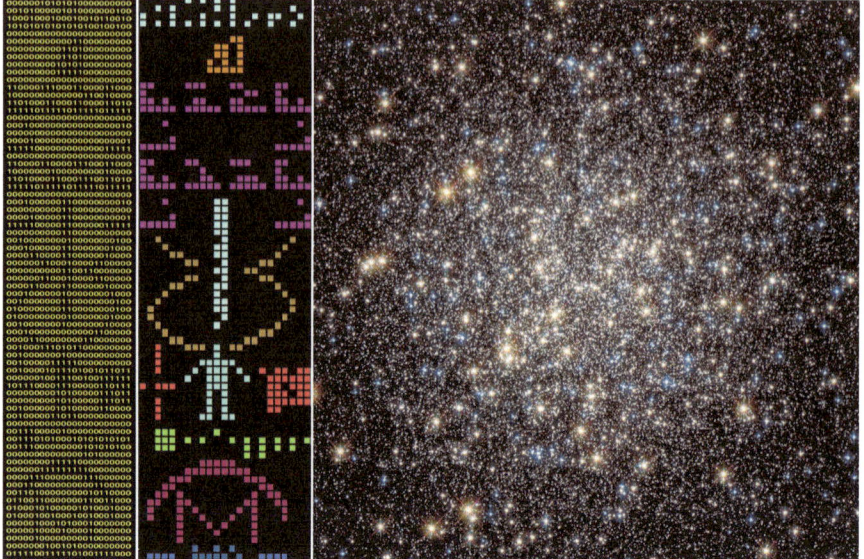

Fig. 7.6 *Left:* The Arecibo message as ones and zeros. When assembled correctly (*center*), it leads to this image. Colors have been added to call out various parts of the graphic. In reality, the message comes only in black and white (or "on" and "off"), as seen in the left frame. (Art © Michael Carroll.) *Right:* The globular cluster known as Messier 13 (M13), target for the Arecibo transmission, lies some 25,100 light-years away in the constellation of Hercules. (HST image courtesy of NASA/STScI/ESA)

rounded by a mesh of copper wire to prevent signals from straying into the nearby antenna. Even in the nearby public café, the microwave ovens are shielded to prevent a rerun of what happened at Parkes Observatory. Green Bank hosts the largest movable land object in the world. Its gracefully curving antenna is taller than the Statue of Liberty. Within the detector, many components must be shielded in gold plating to keep out any interference from Earthly radio sources (Fig. 7.7).

Green Bank's radio dish is the largest movable object in the world, but another radio telescope is so large that it doesn't move at all. In a mountainous region of Puerto Rico, engineers have scooped out a gigantic bowl and inset the vast radio telescope called Arecibo. Built in the 1960s, the radio telescope is managed by several organizations under the umbrella of the National Science Foundation. This titanic radio dish was the generator of the Arecibo message. The dish itself is far too large to steer, so instead, technicians can move the horn—a receiver hanging above it—along cables to aim the telescope.

Arecibo is involved in three major areas of research: (1) radar astronomy, (2) radio astronomy, and (3) atmospheric and ionospheric science. Its second

Fig. 7.7 *Left:* The Green Bank Radio Observatory towers above rural West Virginia, far from any Earthly transmission sources. (Image courtesy of the National Radio Astronomy Observatory/AVI/NSF.) *Right:* The vast array at Arecebo listens in for distant signals. (Image courtesy of the National Science Foundation)

area of interest, radio astronomy, has directly contributed to SETI. Arecibo is the source of data for the Seti@Home project, and has monitored stars for SETI's Project Phoenix (see "Undertakings in the Search for Extraterrestrial Intelligence," below). The huge dish made some of the first radar images of Venus, has imaged near-Earth asteroids, and listens in for the signatures of natural radio sources from the planets to deep space.

High in the Chilean wilderness, the Atacama Desert encompasses some of the driest terrain in the world. Its remote location away from cities, lack of cloud cover, and almost complete absence of radio and light interference, it is the perfect setting for a radio telescope. The Atacama Large Millimeter/Submillimeter Array sits on Atacama's Chanjnantor plateau at an altitude of 5000 m.

ALMA's main array is still under expansion. When complete, it will consist of fifty 12-m antennas, arranged in configurations spread over distances from 150 m to 16 km. ALMA uses interferometry, linking multiple dishes together to create the effect of one large telescope. Its resolution depends not upon the size of each dish but rather on the separation of the dishes and area covered by the entire group. As antennas move farther from each other, their resolution increases. Data from each antenna is combined and processed by computer to emulate the resolution of one giant antenna. In the radio spectrum, ALMA will have a maximum resolution that is even better than the visual resolution of the Hubble Space Telescope. To date, 45 of the 66 antennas are operational (Fig. 7.8).

The backbone of SETI Institute studies in the centimeter region of the radio spectrum is the Allen Telescope Array (ATA) in Hat Creek, California.

Fig. 7.8 The ALMA (Atacama Large Millimeter/Submillimeter Array) telescope consists of 66 dishes arrayed together for higher resolution radio imaging. Erected on the 5000-m high Chajnantor plateau in the heart of Chile's Atacama Desert, the dishes can be moved up to 16 km across the desert floor. (Image © Babak Tafreshi, used by permission)

The array works in similar fashion to ALMA, joining 42 antennas together to search for signals. ATA can carry out simultaneous surveys at multiple wavelengths.

In 2001, Paul Allen, co-founder of Microsoft, stepped up to fund the technological development of ATA, along with the first phase of its implementation. Operations began in October of 2007. The ATA is on the air every day, continuing its search of the heavens.

Calls for Caution

Some people are opposed to any Earthly transmitting altogether, considering it to be dangerous. What if we are advertising our presence to a hostile race, the kind that reared its ugly head in *Independence Day*? SETI researcher Seth Shostak's response is simple: "We've been broadcasting for seventy years. Yes, those signals are weak, but they can be picked up by any society that would have the tech to do us harm. If your worry is motivated by security concerns, it's too late."

The idea that an alien race would come to Earth to steal away our resources does not hold water, either. Water is often the target of invaders. "Take me to your water" could be the theme of TV series or movies such as *V* (1983), David Bouie's film *The Man Who Fell To Earth*, or the Hollywood disaster movie *Battle: LA* (2011). But our Solar System is filled with sources of water. The majority of outer moons are ice balls ready for the taking. They have the advantage of a complete lack of humans crawling all over them. Other resources on our planet are plentiful throughout the cosmos as well. Coming to our Earth to get them would mean dealing with the natives, something that any intelligent being would probably want to avoid.

Undertakings in the Search for Extraterrestrial Intelligence

If sentient beings on Earth are contemplating sending out messages, perhaps others have already done so. This supposition forms the backbone of many projects that have made up the search for extraterrestrial intelligence. The first to apply these ideas in an organized search was Frank Drake. In 1960, Drake had just become the director of the Green Bank National Radio Astronomy Observatory in Green Bank, West Virginia. Drake set up the first system-atic search for extraterrestrial transmissions, calling his project Ozma.[10] In the spring of 1960, Drake put Green Bank's 26-m radio dish to work, spending two months looking for radio signals in the direction of two nearby stars. Drake's targets were Epsilon Eridani and Tao Ceti, both about 12 light-years away. Drake's search didn't turn up any distant relatives, but it brought to bear the disciplined techniques that would lead to a host of radio searches.

In ensuing years, dozens of other projects sprang into being, not only in the United States and the USSR but also in Canada, Australia, France, and Holland. After the initial attempts by Frank Drake and others, a NASA-funded project came together under the moniker of Project Cyclops. The project was tasked with strategy for detecting extraterrestrial transmissions. The final report stated that, "It is vastly less expensive to look for and to send signals than to attempt contact by spaceship or by probes. This conclusion is based not on the present state of our technological prowess but on our present knowledge of physical law." Cyclops recommended chaining a constellation of radio telescopes across the world to search for radio signals among stars within a 100 light-year radius.

---

[10] Ozma was named after the queen of Frank Baum's fabled land of Oz, a remote and unreachable world.

The project was eventually shelved due to cost, but its landmark recommendations continue to inform SETI projects.

Even in the infant stage of Cyclops, researchers recognized that the name of the search game was to survey as many star systems as feasible over as much of the radio dial as possible. These elements form the cornerstone of SETI research to this day. Ideally, a survey would monitor the entire sky all the time at all frequencies at high sensitivity. This daunting task is beyond our current resources, but we are making headway.

A fundamental change since Cyclops is changing the game—the advent of digital electronics. In the relatively primitive days of Cyclops, investigators couched their strategies in terms of such analog approaches as photography and the manual sifting of radio wave records. The digital format enables us to study many frequencies at once. Digital computing enables researchers to swallow wide swaths of the radio dial, breaking it up into tens or hundreds of millions of channels. These signals can be studied with real-time processing, something a Cyclops engineer would never dream of. And with the advent of the digital realm, SETI advances with the same speed that digital electronics accelerates.

In 1981, NASA briefly canceled SETI's government funding. It was reinstated in 1982, thanks to lobbying efforts of Carl Sagan and others. During the funding maelstrom, 1981 saw the practical application of SETI by Harvard physicist Paul Horowitz. Horowitz came up with a device that could analyze 131,000 channels of radio spectrum at one time. His "suitcase SETI" analyzer was ultimately chained to the 26-m Harvard/Smithsonian radio antenna, continuing studies up through 1985.

Hot on its heels was the instigation of Project META, the Megachannel Extraterrestrial Assay. Director Steven Spielberg donated $100,000 to construct an 8-million channel receiver at the Oak Ridge radio telescope in Massachusetts. Project META began its search in 1985. In the ensuing years, the project morphed into Project BETA, which had an increased capacity of a billion channels. At the same time, SETI engineers set up the META II in Argentina to search the southern skies.

A NASA-sponsored project called MOP (Microwave Observing Program) followed in 1992, beginning a systematic study of 800 nearby stars at about the same time. MOP was designed to utilize the massive radio antennas of NASA's Deep Space Network, along with the 43-m radio telescope at Green Bank and the 300-m radio antenna at Arecibo in Puerto Rico. Although Congress canceled MOP's funding after just one year of operation, private SETI supporters continued funding its operation under the name Project Phoenix. The resurrected program made use of Australia's Parkes radio telescope, along with the instruments in Puerto Rico and West Virginia.

In 1993, the SETI landscape was significantly altered when the U. S. Congress zeroed out all SETI funding. Now, financial backing had to come from various private endowments and public advocacy groups such as the Planetary Society, which has always been at the forefront of SETI studies. The SETI Research Institute in Mountain View, California, continues the search with observatories across the globe, including the Australia's Murchison Widefield Array, the United Kingdom's Lovell Telescope, and the Low Frequency Array located in the Netherlands.

The data from various SETI investigations amount to vast quantities of information. Even in the digital world, sifting through that pile is one of the greatest challenges facing the search today. One clever strategy insti- gated by the Planetary Society and the University of California at Berkeley is called SETI@home. Over 3 million individuals use their private comput- ers to participate. SETI@home's computer software acts as a screen-saver while at the same time analyzing raw data from SETI facilities. The data is processed and automatically returned to SETI teams where it can be further scrutinized. SETI@home is now the largest computing experiment in the world.

Radio waves are only part of the electromagnetic spectrum. If your com- puter or cable TV gets data through fiber optic cable, then it is obtaining its data stream from infrared or visible light. Light has an advantage over radio in that it can be focused into a concentrated beam, as in a laser, to communicate in more directed ways. Although visible light fades as it hits clouds of interstel- lar dust, light in the infrared will pass through these obstacles. Pulses of laser light could make up messages just as radio waves do. Optical SETI searches are underway, looking for laser light "pings" from other star systems. The 1-m Nickel Telescope at the Lick Observatory is involved in Optical SETI. It is sensitive enough that the instrument could discern an optical signal similar to our own transmitters from 500 light-years distant.

The SETI search has gotten a boost from famed physicist Steven Hawking and billionaire Yuri Milner. The two have invested $100 million to start a new SETI initiative called Breakthrough Listen. The 10-year project is touted as the most powerful search for extraterrestrial life in history. Breakthrough Listen will monitor the million stars closest to Earth.

As we listen for messages from other worlds, we are faced with the question of how to recognize them. In Carl Sagan's novel *Contact*, an alien race simpli- fies things for us by broadcasting back one of our own television programs before they try to send us something with a little more meaning. But this strategy would only work within a distance of 50 light-years from us, as that is how far our old TV transmissions have traveled. But extraterrestrials have

Fig. 7.9 Our radio telescopes may not be the only instruments searching for life in the cosmos. Physics may dictate similar engineering designs throughout the galaxy. (Art © Michael Carroll)

other options to help us. Repetition of form would be the first recognizable earmark of intelligent design in a transmission (Fig. 7.9).

Jill Tarter offers another reason that we may not have yet recognized alien transmissions. They may be disguised to keep us in the dark until we reach a certain level of sophistication. Radio astronomers now chart two general classes of rapid radio bursts. The first type is repetitious, usually from pulsars or magnetars. The second type may emanate from galaxies outside our own, perhaps as a product of colliding neutron stars. The latter class is the one that intrigues Tarter. "The fast radio bursts with dispersion, which seems much greater than any number of electrons you could get in any line of sight in the galaxy, has led people to conclude that these are extremely powerful extragalactic sources. But a couple of people, including me, have suggested that we could engineer a signal like that in the lab. We could have made it look like something that could get caught in a pulsar survey and appear anomalous and worth taking a second look at. There have been a number of suggestions about creating signals that look almost natural, so that when we, the youngsters in this game, build equipment to look at the universe and observe, they'll get caught."

These types of radio bursts may not be that far away after all. Of the many signals logged so far, a scant seventeen fall into a category that may be artificial signals such as the one Tartar envisions. "They would end up in our databases and some graduate student or post doc will say, 'Hey, that pulsar had that period last month, and two weeks later it had this different period and today its back to the first period That's pretty strange.'" Those "almost natural" signals are the ones that the astronomers are going to trip over. It is the engineered signals—the ones that don't do what nature can do—that are the types of signals for which SETI will be watching.

A Different Kind of Message

It may be that civilizations broadcast conventional radio or optical transmissions for only a brief period before they figure out a better way. Put another way, although the signals are out there, we may have no way of detecting them with our current technology. In a sense, there is historic precedent for this idea. If we remember the strategies offered by nineteenth century observers for communicating with the planets, those strategies seem naïve by today's standards. Perhaps the kinds of SETI exploration we do today will seem naïve a hundred years from now. That may depend not so much on technological breakthroughs as on what we learn in the realm of physics. The horizon may hold breakthroughs for faster-than-light travel and faster-than-light communication. But waiting for such breakthroughs doesn't make sense to Seth Shostak. "It sounds to me a weak argument to say 'you guys are wasting your time because you don't understand hyper-dimensional physics' (whatever that turns out to be). We don't, so we do what we can now. Maybe this is the wrong approach, but you can always make an argument that says 'Let's sit on our hands and wait for something better.'" SETI's operating assumption must be that there are civilizations out there that are either actively seeking contact, or they don't care if their transmissions are discovered by others.

A tiger team of world experts in astronomy, engineering and information technology convened in 1997 to help SETI strategize its future approach to the search for evidence of life among the stars. Their three-year study, christened SETI 2020, resulted in several overarching recommendations, Jill Tarter explained. "SETI 2020 suggested we do three things. It suggested that we begin optical SETI because the fast rise-time photodiodes had finally come out from under wraps in the military and were now available and affordable. We have started optical SETI in a couple programs, from Harvard to Panama to Australia."

The committee's second recommendation was that SETI should build its own telescope and build it as a large number of small dishes. SETI has the computing power needed to slave together multiple dishes, and the approach is the way to beat the cost of building a large telescope. SETI started the ATA observatory, and hopes to eventually expand the antennas up to a grand total of 350.

The final recommendation from SETI 2020 was to build an omni-directional sky survey, Tarter says. "The reason is that the way radio telescopes observe, the signal has to be persistent. You have to be able to go back and find it. If it's transient, none of us are very sensitive to that. We've seen the fast radio bursts that have survived the false alarms, and statistics say that there are probably 10,000 of those going off in the sky every day, but they're really hard to find because they're only a millisecond, and you have to be looking at the right place. It could as well be that for SETI signals there would be these transients."

Tarter imagines a distant civilization with a long list of 100 million stars that they are going to ping periodically. They move their transmitter from one target star to another, but when the signal finally falls upon Earth, it is unlikely that our observers would be looking in the right place. The only way to circumvent the problem is to look at the entire sky. "To get any kind of sensitivity, you need a lot of computing power," Tarter says. "We are actively engaged here with try-ing to figure out a way to do the same thing in the optical, with using small, amateur kinds of telescopes to get large fields of view, with computing that separates out things to give you a decipherable signal for these transient flashes." In the meantime, we wait and we watch and we search.

However, if we are successful, if our search turns up an intelligent species desiring to communicate with us, we are faced with a new challenge—what would we say to them? Who would decide?

How would we proceed if we could finally tease an artificial signal from the song of the universe? Any possible detection of a signal from another star system would need to be carefully verified. Observatories around the globe would monitor the source for repeating signals. Its pattern, frequency and nature might be evidence of intelligence, but that intelligence might be on Earth. Analysts would need to confirm that the transmission is not interfer-ence from a terrestrial source or some type of equipment malfunction. Once all these steps have been taken, in the course of weeks or even months, official news would be made public (although rumors would have probably leaked out by then). This kind of discovery is fraught with heavy social and cul-tural implications, so its announcement must be made carefully so that all cultures and language groups on Earth will understand it without fear or misconceptions.

Because of the delicate nature of such an event, SETI researchers have agreed on a procedure. The protocol is called the Declaration of Principles

Concerning Activities Following the Detection of Extraterrestrial Intelligence. The plan is a sort of road map outlining an agreed-upon course of action. Astronomers and governments around the world would be notified directly. No attempt at secrecy would be made, as this discovery should be celebrated by all people on the planet. A response would be made in such a way as to represent a consensus of the population at large (and not preferentially by the discoverers). Odds are that the signal we receive is from a civilization far older than ours. It may be that we can never decipher the message, that it is beyond our understanding. It might also be that the message was intended for another advanced civilization, and we are inadvertently eavesdropping on a message we can never hope to decipher. Still, the very event would show us that we are not alone, and that technological civilizations such as ours can grow, flourish and mature into older age.

Our outgoing message would need to be more sophisticated than what has come before. The Arecibo message and the Pioneer plaque all carried the equivalent of a few thousand bits of information, comparable to a few paragraphs of English text. The improved Voyager records offered a few million. The technique of radio transmission offers the possibility of far more data. Our message would need to begin with common ground. This is difficult, as we would have no idea what life forms were on the other end, how they related to their physical environment, or what they thought of the universe around them. But we would share one thing in common—science. To build a radio antenna, alien beings would need to have stepped through the same stages of technological development that we have, involving physics, mathematics and even chemistry and biochemistry. If they are searching the cosmos, then they also share knowledge of stars, planets and galaxies. Perhaps we could find enough commonality among these subjects to make meaningful contact.

Another approach is to simply inundate them with enough data that they can see patterns, recognize trends in our information, and decipher our photos, music, and other messages. One strategy would be to send the entire contents of the Internet. This concept, put forth by Seth Shostak, is a sort of 'hail Mary' toss of the radio football. The idea is that if we are going to send something to a civilization a hundred or even a thousand light-years away, we might as well send everything. The conversation will be one-sided because of the great distance and time lag involved. "I just worked out what sorts of things you could send," says Shostak, "for example, the Library of Congress or all of the world's music. My personal opinion is you just send the entire Internet. There's a lot of redundancy in all that." That redundancy will provide clues to the recipient, patterns and keys to the nature of our communication. The images, text, and audio files of the world's Internet should afford enough information to begin decoding our transmission, Shostak feels. His precedent is the case of Egyptian hieroglyphics. The ancient painted symbols of the pharaohs were so ubiquitous throughout the temples and monuments that it was only a matter of time until a brilliant linguist came along to decipher them.[11] The petabytes of material from the Internet should enable our recipients to make enough correlations to figure us out on at least a basic level.

If I Could Talk With the Animals...

The process of communicating with an alien species is a prodigious challenge, but we have already made inroads in interacting with a community that has grown up in a microgravity environment, a species that possesses neither written language nor pictorial traditions, populations of creatures who spend most of their time in an environment devoid of gaseous oxygen. They communicate most often not by spoken language but by sonar. Unlike us visually-oriented humans, they are auditory beings. They are the cetaceans.

Dorian Houser is an acoustical researcher at the National Marine Mammal Foundation (NMMF) in San Diego, California. For several decades, Houser has been working with *Terciops truncates*, the bottlenose dolphin. He is a foremost expert in dolphin behavior and communication. Dolphins, Houser says, are quite alien beings in their own right, which complicates our communication with them. "We have evolved in very different worlds, so while we face some of the same challenges (reproduction, food, the ability to protect ourselves), the three-dimensional world in which they live is so very different from ours. Cognitively, the question becomes, 'How have they been wired to deal with that environment differently than we've been wired to deal with ours?'" Dolphins are clearly intelligent. They solve complex problems, come up with creative solutions to puzzles, and engage in advanced social behavior. They have a brain to body mass ratio second only to humans, but a large portion of that brain is probably due to the fact that they have to deal with sound as their primary physical characteristic for engaging with their environment. The processing of all that acoustical data takes physical space. Houser contrasts this to human brain architecture: "Even though we do a lot of verbal communication, we also do visual communication: body language and the like. Dolphins probably have postural communication as well."

---

[11] That linguist was French scholar Francois Champollion. He had help from the Rosetta Stone, a three-language tablet that held three parallel decrees by Ptolemy V. Champollion's breakthrough discovery marked the beginning of modern Egyptology.

The initial attempts at communication were on a fundamental level. Houser looked for commonality between the human and cetacean species, finding food, reproduction and other similar challenges that might drive certain cognitive processes. With the dolphin, marine researchers go to the lowest common denominator first. "We establish our communication around food. It's the most basic, simple level of communication. 'We would like to give you this.' And then they respond. We are doing operant conditioning, but over time, when you work with certain individual animals for very many years, they show their intelligence by their ability to pick up novel concepts very quickly or to 'game.'" In other words, the dolphins toy with their trainers. Houser and his colleagues recently did a study in which dolphins produced echolocation clicks. The scientists studied the dolphins' ability to detect changes in a target location. The researchers tried to mask the echolocation clicks with other sounds, so the dolphins could not hear their own clicks. But very quickly, the dolphins began moving the frequency of their echolocation clicks around. They anticipated what the scientists were planning on doing. "The animals play games with us," Houser says. "They're certainly figuring out things on their own. They are intelligent enough to be able to adapt to a complex situation."

Dolphins are equipped to interact with an environment alien to humans, and they're good at it. But attempts to replicate dolphin language acoustically have met with little success, says fellow NMMF engineer Alan Lewis. "The problem is that you have to put out a nice clear signal, but it's a signal that the dolphin can change as they go. 'Oh, now I'm looking for something buried, and I'm going to reduce the frequency just on the fly.' They switch frequency and use multiple frequencies and then it's like, 'Oh, I heard something. Maybe I'll turn my head 30 degrees this way and ping it again.' The dolphin is so highly mobile. In engineering terms, a dolphin is a remarkable Autonomous Mobile Broadband Sonar."

As for communicating with an alien civilization, Dorian Houser cautions that humans tend to imprint anthropomorphic features into their activities. "Part of the problem is that we come in with a fairly human bias. For example, we assume that a word equates to object or action. That may or may not be so. Dolphins have different dialects, but we've also seen when animals move in with another group of animals, a group that they're unfamiliar with, they will adapt, they'll hybridize their signal. We need to take that into consideration and be fairly fluid. The other thing we need to take into consideration is that even in our own study of language, we come across words that you can't really translate. We hear a particular signal at a certain time, and we think it correlates with a certain behavior or mood."

One prime example of an archetypal dolphin utterance has to do with distress. Dolphins make a stereotypical sound if they appear to be distressed. But interpreting that signal is difficult: does it mean the animal is in distress, or is it warning others that there's something here to avoid?

The pitfall, Hauser asserts, is that in communicating with different species, we come at it with our own bias. "As a biologist, the number one thing you need to do is to understand the natural history and the behavior. Unless you understand how this animal makes a living and interacts with its environment, and interacts with other individuals of the species, it's difficult to know where you would start."

Pondering the human bias question, Seth Shostak says, "It's pretty hard to respond to that, because any response you give is probably going to be anthropocentric, unless you ask the dolphins…I'm willing to accept that dolphins have a very high encephalization quotient, they have a big brain, and that generally means intelligence. On the other hand, they don't write great literature, they can't pick up a soldering iron and manipulate it very well, and they haven't built radio transmitters. From the standpoint of SETI, they may be very interesting creatures, they may have subtle behaviors, but they're not intelligent by our very simple definition."

Even biologists have difficulty defining the range of intelligence in the animal kingdom. Behaviorists and psychologists have studied problem solving, memory, tool use and communication in various species. Studies have included mammals such as the cetaceans and pinnipeds, elephants and primates (chimps and apes). Other fields of research have included birds (parrots, ravens, magpies), reptiles, and cephalopods. Chimpanzees do not use trial and error to harvest out-of-reach bananas, for example, but rather use tools or elements in their environment, such as stacking boxes or moving other objects to bring their goal into reach. Octopi in the laboratory setting have been observed throwing rocks to break other aquariums, squirting water onto lights, and unscrewing the lids of jars to obtain food. Complex behaviors such as these blur the lines between learned behaviors and intelligence.

Blurring those lines even farther is the issue of language usage, especially in primates and apes. These close cousins of humans develop complex social structure and need to communicate with each other frequently. Primates depend strongly on body language, while humans are more dependent on vocalizations and complex sounds.

Linguists tell us that the languages we humans use have two unique characteristics: they are both 'open' and 'discreet.' The quality of 'openness' relates to our ability to communicate about new situations, objects and ideas. Chimps and apes appear to communicate nearly entirely in the realm of present emotions and goals. They live wholly in the present. The discreet nature of language refers to two properties of language: (1) continuous shifts in the acoustic signal lead to distinct sounds (intonation lends meaning), and (2) discreet units result in changes in the content or meaning of the message (like words).

Physically, primates are at a disadvantage for the spoken word. They do not have the vocal cord and mouth construction to make the sounds involved in human verbal language. The human tongue and larynx are more flexible than those of primates. But some research, ongoing since the 1960s, strongly suggests that at least some individual primates can learn to use the American Sign Language for the Deaf (ASL). The question is: Do primates have the mental capacity to understand and use symbolic communications, and to use them creatively? Evidence now indicates that some African apes have learned to use a simplified version of ASL. Researchers differ on just how deep their understanding and communication skills go. Can they combine linguistic elements to form new words and concepts? Can they speak and conceptualize in abstract ways?

Washoe, a West African chimpanzee, was the first research subject. University of Nevada psychologists Allen and Beatrix Gardner began studies of Washoe in the 1960s. Over the course of her life, the chimp learned more than 250 signs, and purportedly combined signs to create novel phrases such as "You me go out hurry" and "Give me sweet." Researchers Roger and Deborah Fouts claimed that she signed "water bird" after seeing a swan. Skeptics are unconvinced of Washoe's impromptu word inventions, averring that the chimpanzee was merely signing what she saw: water and a bird.

Over the decade of the sixties, the field of human/ape sign language thrived, increasing supporters and skeptics alike. Apes learned to use plastic symbols or keyboards to communicate with observers, and several learned elements of ASL. One of the most famous stars in ASL/primate language is the African lowland gorilla Koko. Her training in sign language began in 1972 at Stanford University, under the care of Francine "Penny" Patterson. Patterson chose ASL because of other researchers' earlier successes with chimpanzees. Those who study primates have established the fact that gorillas have a natural gestural language in the wild and in zoos, utilizing consistent signs to communicate universal concepts. According to Patterson, within a matter of weeks, both Koko and another ape were not only using ASL but were combining elements to create new words. According to Patterson and other scientists at the Gorilla Foundation, Koko "knows 1000 signs and understands spoken English." But critics remain skeptical, citing a lack of published and peer-reviewed studies. One outspoken critic is psychologist Steven Pinker. Declaring that no primate has learned sign language, he asserts that sign language is more than a system

Fig. 7.10 A few of the semi-intelligent creatures with whom we share our planet. Studying these species may help to hone our skills for communicating with alien species. *Left:* Lowland gorilla; *Right:* Bottlenose dolphin. (Gorilla: Pierre Fidenci, Wikipedia commons: https://commons.wikimedia.org/wiki/File:Gorilla_gorilla11. jpg; dolphin image courtesy of NASA)

of gestures. Pinker contends that primates have not grasped the full, systematic grammar of true linguistic communication (Fig. 7.10).

Perhaps our definition of intelligence is flawed. It may be that humans do not yet comprehend the depths and nuances of animal behavior, nor other species' interaction with—and perception of—their environment. We may simply be too anthropocentric in our evaluations. But one thing is clear: if we use the SETI definition for intelligent life, even the most advanced primates do not qualify. They will not be building any large antennas in the near future.

Assumptions

Dorian Houser agrees with many SETI experts who suggest that if we encounter sentient beings in the cosmos, we must operate on one assumption— that there are basic shared principles between us and them. The acquisition of energy is the most fundamental. If an alien species is not an intelligent photosynthetic race, they probably have to eat. Assuming that the extraterrestrial beings have a visual modality, Houser asks, "would we send images or video that shows eating, accompanied by some sort of sound recording? I can't assume that all beings have sexual reproduction, so sending pornographic materials would probably not be great. There are species on Earth that are asexual. You have to look for something that has a common basis among all living organisms. Assuming this being is intelligent and has similar sensory modalities, you'd have to utilize those modalities using basic descriptors that they can determine by their sensors. You have to find a common point from which to build other points of communication. You have to build some kind of common linkage." This, in fact, is exactly what researchers such as Frank Drake and Seth Shostak have tried to envision.

As discussed earlier, another possible answer to Fermi's great question is that other races simply do not want to be seen. Any wildlife field biologist will tell you that to get good data, you must camouflage yourself so as not to disturb the target of study. Animal behavior changes in the presence of a human observer. In the case of some rare and endangered creatures, keepers must disguise themselves in order to feed and care for the young, in this way preparing them for life in the wilderness. Perhaps aliens are exercising the same such caution. Either they are studying our progress as a species or they are watching us for their amusement (and one must admit, the human race is often amusing). This is called the "zoo hypothesis."

However, some SETI experts believe this is an anthropocentric view. If we can easily see galaxies billions of miles away, and can detect the radio noise from the Big Bang some 13 billion years ago, they ask, how would aliens manage to shield from us the evidence of their existence?

There are other objections to this theory. Only the closest star systems could know of the presence of *Homo sapiens.* The only radio evidence of our existence has come about since the Second World War in the form of high frequency radio, TV and radar. Would an advanced race turn Earth into a zoo before they knew about human civilization? After all, entomologist E. O. Wilson studies ants in his shirtsleeves, without the aid of an Ant Man disguise.

The zoo hypothesis elicits a response from Seth Shostak, and it's a response shared by some other SETI researchers. "The zoo hypothesis is kind of 'cagey.' It seems like a lot of effort for something that I can't understand the motivation for. We don't present a threat to anybody. I just can't imagine some Klingon saying, "Well you know, there might be trilobites on that planet. We'd better hide ourselves from them.""

UFOs

If extraterrestrial zoologists are visiting us, perhaps humans have glimpsed them. The phenomenon of UFOs (Unidentified Flying Objects) has captured the imagination of many. If the galaxy is filled with alien life, surely our planet has been visited once or twice by advanced spacecraft, perhaps even craft piloted by the aliens themselves. Arguments supporting alien visitation are widespread. Among the general populace worldwide, witnesses include trained pilots, military observers, and a host of field scientists.

The most famous alien encounter claim comes from the case of an alleged UFO crash-landing in Roswell, New Mexico. The incident took place in the summer of 1947. Idaho businessman and pilot Kenneth Arnold spotted what he described to local papers as crescent-shaped objects darting above the mountains "like saucers skipping in water."[12] The same month, a rancher 100 km north of Roswell, New Mexico, discovered some suspicious-looking debris on his land. He called the local military authorities at nearby Roswell Army Air Field. The RAAF dispatched a team that removed the debris, officially calling it a "flying disk." A day later, a formal military press release called the wreckage a "weather balloon." It would take nearly fifty years for the truth to come out. The debris was no weather balloon, but it wasn't an alien spaceship either. The wreckage was part of a top-secret Cold War project called Project Mogul. Mogul was a classified attempt to monitor Soviet nuclear testing by lofting microphones on high-altitude balloons. The balloons were made of polyethylene and filled with helium. On June 4, 1947, just four days before the Roswell incident, the RAAF launched two dozen balloons tethered together to form a high-altitude line of stations. Later studies indicated that winds at the time of deployment would have carried any wayward Mogul balloons directly toward the ranch where the "crashed spaceship" was discovered. The change in official air force position added fuel to those who thought a cover-up was under way (Fig. 7.11).

Roswell is one in a long line of reports attributed to alien visitations. Others involve blinking lights, movement that seems to defy the laws of physics, burned landing sites and even alien abductions and cattle mutilations.

Several studies worldwide have attempted to explain UFO-related phenomena. Official independent investigations are typically able to discount 94 % of all reports studied. But what about that last 6 %? The skeptics reason that there are many causes for disbelief.

First, the skeptics point out that for decades Earth has been under orbital surveillance by our own satellites. Our orbiting spacecraft chart the weather, monitor the health of vegetation, and map ocean currents. Our military satellites are specifically designed to see artificial craft in and above the atmosphere. None of these busy satellites have seen any flying saucers. Our airports and armed services operate sophisticated radar constantly, but with no definitive revelations of cosmic visitors. The North American Air and Space Command in Colorado Springs in the United States carefully monitors thousands of chunks of space debris, some of it as small as a baseball. Their sensitive equipment has found nothing alien over the course of nearly five decades.

---

[12] It was a misquotation of his report that led to the term "flying saucers."

Roswell Daily Record

RECORD PHONES
Business Office 2288
News Department
2287

Claims Army Is Slacking Courts Martial

Indiana Senator Lays Protest Before Patterson

RAAF Captures Flying Saucer On Ranch in Roswell Region

House Passes Tax Slash by Large Margin

Defeat Amendment By Demos to Remove Many from Rolls

Security Council Paves Way to Talks On Arms Reductions

No Details of Flying Disk Are Revealed

Roswell Hardware Man and Wife Report Disk Seen

Ex-King Carol Weds Mme. Lupescu

Satellites

Fig. 7.11 Local New Mexico newspapers covered the famous Roswell, New Mexico incident (*top*). The day after this article appeared, the air force printed a "correction" of the story, which was, in fact, designed to disguise an ongoing secret military project called Mogul, seen below just before launch from Holloman Air Force Base in Alamogordo, NM. Mogul was designed to monitor Soviet nuclear tests using microphones lofted by balloons. (*Image source*: STRATOCAT, http://stratocat.com.ar/bases/30e.htm.) (Article source, https://commons.wikimedia.org/wiki/File:RoswellDailyRecordJuly8,1947.jpg)

However, there is another phenomenon that some claim points to alien visitation: crop circles. Across the countryside of England in the 1970s, reports poured in of bizarre patterns impressed upon fields of wheat and rye. The dramatic shapes were geometric, and so regular that many claimed they could not possibly be due to natural causes. Scientists monitored the sites to see if wind could have given rise to such elegant vegetative shapes. Crop circles have materialized at many sites throughout the world. British observers have noted that they tend to appear near population centers or roadways where their locations afford easy access. Crop circles gained even more fame in the late 1980s as the story of formations in Hampshire and Wiltshire spread through the general press and TV. In 1991, *The New York Times*[13] reported that two British men had made an amazing claim: they were the ones who executed the massive hoax. David Chorley and Doug Bower followed tractor rows into the center of the fields, and from there traced the patterns, trampling down the wheat using "two wooden boards, a piece of string and a bizarre sighting device attached to a baseball cap."

Advocates of crop circles remain dubious about the two men's claims, despite their demonstration of their techniques. They say that crop circles are too widespread and universal to have been instigated by a couple of hoaxers. Crop circle skeptics have pointed out that a field of cereal as a mode of communication seems inadequate to those who have mastered relativistic space travel. As the authors of the book *Cosmic Company*[14] put it, "It seems bizarre in the extreme to think that sophisticated extraterrestrials would go to so much trouble to tell us so little."

As for Earth's visitation by advanced alien cultures in disk-shaped spacecraft, Carl Sagan said that, "Extraordinary claims require extraordinary evidence." If this is the case, many analysts maintain, then UFOs have failed the test. Unless, of course, flying saucers are piloted by alien zoologists.

Civilization's Fingerprints Here and Beyond

Extraterrestrial beings may have left calling cards unintentionally, littering the cosmos with their discarded garbage. When studying ancient civilizations, archeologists always look for the local dump. The detritus of a civilization can tell us much about habits and lifestyles. For example, in the 2009 digging season, archeologists excavating a tenth-century B.C. refuse dump adjacent to a royal building near Jerusalem's Temple Mount discovered the first in situ clay seal impression of a Judean king. The 9.7 x 8.6 mm oval impression mentions King Hezekiah (727–698 B.C.), and displays a winged Sun symbol.

---

[13] *The New York Times*, "2 Jovial Men Demystify Those Crop Circles in Britain" by William E. Schmidt, September 10, 1991.

[14] Cosmic Company by Seth Shostak and Alex Barnett, Cambridge, 2003.

The scientists found 33 other clay seals, some still bearing impressions of the cords that they sealed over the tops of food storage jars.[15] One person's trash is an archeologist's treasure.

We humans, as a race, are leaving behind bits of trash that will last far beyond us. Rather than elegant royal inscriptions, our artifacts include aluminum cans, Styrofoam cups, and various metals and plastics. Our spacecraft will also bear silent witness to our presence. We've left many landers and surface probes scattered across the Solar System, many in places where there is little or no weather. These range from Soviet *Venera* Venus landers to multinational rovers and landers on the Moon and Mars. We've even deposited landers on asteroids, comets, and Saturn's moon Titan.[16] Perhaps future travelers will stumble upon comparable litter in other planetary systems, or even in our own. It is possible that a curious race has left interstellar robotic spacecraft or landing probes in any one of myriad locations right in our own Solar System.

Some advanced civilizations may have left remnants of themselves in gigantic assemblies the size of entire solar systems. Even if an alien race has come and gone, cosmic archeologists might be able to discover remnants of such civilizations. There is good reason to suspect that mega structures have, in fact, been erected somewhere in the cosmos. The history of humanity is a history of expansion. Biology teaches us that a species grows and expands, or it atrophies and dies. Human civilization has expanded across the globe, leaving many artificial structures and examples of landscape engineering in its wake. If the trend continues, our species will one day move out into the Solar System, perhaps building vast energy-catching nets in globes around the Sun, or building other super-structures linking our world to others.

Our search for cultures among the stars can be informed by projecting the course and expansion of civilization in our own Solar System. Futurist Nikos Prantzos[17] submits that in the future, three categories of civilization may develop around Sol. The first and most familiar to us would inhabit planets and moons nearby. Strategies would be involved. The first is to erect artificial biomes such as habitats, pressurized tented craters and valleys, or vast domed cities above or underground. A second strategy, limited to one or two planets, would be terraforming—engineering the environment of an entire planet.

---

[15] Before the invention of locks, goods were placed in jars with cloth covering the top. Around this cloth, a rope tied the cloth in place. A clay seal was then placed on the knot so that the goods could not be tampered with unless the seal was broken.

[16] In the case of Venus and Titan, weather will corrode our little probes fairly quickly, and only slightly less so on Mars. But in the airless environments of the Moon, asteroids and comets, our artifacts may last for billions of years.

[17] For more, see *Our Cosmic Future: Humanity's Fate in the Universe* by Nikos Prantzos (Cambridge University Press, 2000, translation by Stephen Lyle).

The second class of civilization, according to Prantzos, would settle the asteroids and comets. Missions like Rosetta and Dawn have proven that these small cosmic bodies are rich in minerals and other resources—including water—for fabricating an independent society.

Prantzos envisions a third brand of civilization at an apex of the developmental scale. This culture, far more independent in nature, would venture beyond their own system, expanding across interstellar space in generation ships that carry their own self-sustaining environments.

It would be Prantzos' second category of civilization that would stay home and construct vast mega structures throughout their planetary system. The concept of a Sun-surrounding energy sphere was popularized in a 1960 paper[18] by physicist Freeman Dyson and is often referred to as a Dyson sphere or Dyson bubble. Dyson doubted that any advanced extraterrestrials would invest much time in sending messages across space, Arecibo style. But he did think that advanced beings might alter their interplanetary neighborhood to such a magnitude that their modifications would be detectable, even across astronomical expanses. Dyson's "sphere" would intercept the energy from its star for use by civilizations on one or more planets within it. This vast globe, billions of times the size of an Earth-sized planet, could encompass enough space to accommodate millions of years of population growth. The infrared radiation (heat) of a sphere that huge would be easily visible to Earth-bound telescopes, even from hundreds of light-years away (Fig. 7.12).

It was an old idea with a new spin. Konstantin Tsiolkovsky's 1895 book *Dreams of Earth and Sky* described a technologically advanced society extending its activity into interplanetary space, capturing its sun's energy. In 1937, author Olaf Stapledon's classic science fiction story *Star Maker* described a host of advanced alien cultures: "Every solar system was now surrounded by a gauze of light traps, which focused the escaping solar energy for intelligent use."

Writer Larry Niven envisioned a similar arrangement using a ring the size of Earth's orbit (300 million km across). Other variations are undoubtedly possible, constructions large enough for prying terrestrial eyes to discern within distant star systems. Other mega structures we might see include space elevators linking planets to moons or great webs of orbiting tethers used for launching payloads or harvesting energy from a planet's magnetosphere.

Another type of super-structure is called an O'Neill colony. Princeton physicist Gerard K. O'Neill popularized the idea in his 1976 book *The High Frontier: Human Colonies in Space.* O'Neill envisioned a pair of titanic cylinders spanning a total of 50 km in length and 5 km in diameter. For

---

[18] "Search for Artificial Stellar Sources of Infrared Radiation" by Freeman Dyson, *Science*, 1960·

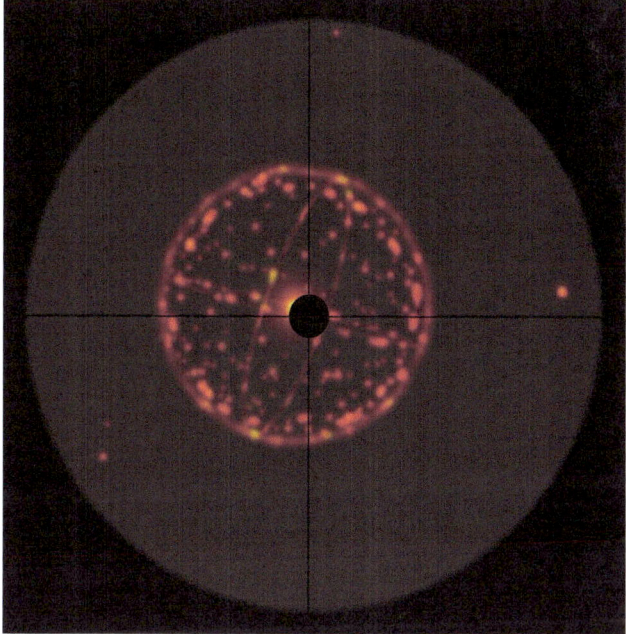

Fig. 7.12 If our infrared (heat-sensitive) telescopes could resolve a solar system surrounded by a Dyson sphere, the view might appear something like this. (Art © Michael Carroll)

stability, the two cylinders—mounted end to end—would counter-rotate, keeping the long axis pointed at the Sun. Each cylindrical habitat would have three landmasses running its length, with clear window panels paralleling them in between.

O'Neill's design incorporated mirrors that would open and close to create a day/night cycle for its inhabitants. The hinges of each mirror would affix to the ends of the cylinders farthest from the Sun, while the open ends would shunt sunlight down onto the world inside the cylinder. The slow spin of the cylinders would create simulated gravity so that the landscapes would cling to the inside walls. Soil and rock in the land areas would be transported from asteroids or the Moon. Land areas would cover 500 square km, enough space to accommodate several million people. Over this land, engineers could construct landscapes with low-lying mountains, waterfalls, lakes and rivers. Forested regions would play host to wildlife, while mountaineers could jump from peak to peak in the lower gravity near the cylinder's center. If an alien civilization fabricated such structures, they might be detectable even from light-years away (Fig. 7.13).

Fig. 7.13 Comets such as 103P/Hartley may one day provide raw materials for our civilization as we migrate throughout our Solar System. Has such a scenario taken place near Earths of distant suns? (EPOXI mission image courtesy of NASA)

Impressive as O'Neill's colony cylinders would be, designers have visualized even larger colonies. Moderate to mega asteroids could be hollowed out to encompass living areas. Vast rotating rings could house entire biomes with cities, wilderness areas, and seas. Hoops could be erected around an entire planet, forming a toroidal rotating donut-world encircling the host planet. Such rings could also encircle a star, taking as much space as the orbit of a planet. The largest of these mega structures could be detected at great distances, either reflecting the light of their host star or giving off copious amounts of energy on their own.

Several researchers suggest that we may have already found a superstructure located where the star KIC 8462852 is, informally known as "Tabby's star." The Kepler space telescope first detected a series of drops in light levels that followed no familiar patterns. Researchers first proposed that a dust cloud or a flock of comets was blocking the starlight. When astronomers looked back at the historical record, they realized that the star has been dimming over the past century, a baffling phenomenon. If the dips had been due to transiting planets, patterns would have shown periodic repetition. But the star's light has been changing erratically, at times dipping by as much as 20 % in a 4-year study period. For comets to cause the starlight shifts, the light levels would demand some 648,000 comets, each 200 km across, passing across the face of the star, some researchers contend. These numbers, they say, make the comet theory implausible. Is it possible that the light of Tabby's star is being obscured by an extraterrestrial construction project? Skeptics of the alien superstructure idea point out that it would be difficult to construct an object covering a fifth of the star in just a century. Additionally, the mega structure would likely radiate infrared radiation (heat), but the IR signal coming from Tabby's star appears to be normal. For now, the bizarre behavior of Tabby's star remains a mystery.

The advanced level of societies that could enshroud entire planetary systems is hard to imagine. In 1964, Soviet astronomer Nicolai Kardashev devised a scale to describe how far technological civilizations had advanced. His scale is based on the amount of energy a given civilization uses, and how much of that energy is invested in interstellar communication. His scale breaks down extra-terrestrial civilizations into three categories. A Type I society has the capacity to use and store the energy from its primary star that falls upon its planet. Our own civilization is close to being a Type I. The light from our cities and the output of our energy plants currently put out a meager amount of energy, and would be difficult to detect from even the nearest stars. Popular theoretical physicist Michio Kaku projects that human society will reach full Type I status within the next two centuries, and Type II (see next) within a few thousand.

A Type II Kardashev civilization is capable of channeling the energy of its entire star, as would a culture constructing a Dyson sphere. Type II beings harness all the energy of their sun, transferring it to their home world. Their energy footprint would be evident even at interstellar distances.

Kardashev's Type III civilization is able to control the energy of the entire galaxy. Such an advanced technological society is so far outside of our techno-logical experience that they are difficult for us to even envision. Recognizing such a civilization would be nearly impossible with our current technology. In fact, we may even be submerged in one right now, ignorant of their usage of the energy of even our own Sun.

SETI's Jill Tarter points out that the WISE spacecraft (Widefield Infrared Survey Explorer) has contributed to the search for advanced technological fingerprints. "Jason Wright and collaborators at Penn State used the WISE data in the search. First, they went looking for Kardashev III civilizations, big cosmic miracles. They didn't find any, but they found 93 sources that look a bit strange. Now, they are going through to do the much harder job, which is looking for the point source, akin to a Dyson sphere." Other projects are under consideration as well. In Hawaii, a team led by Jeff Coon proposes building the ultimate detector. They call it Colossus. The giant radio telescope would span 74 m across. An array of small optical dishes, phased together, would be incredibly powerful, Tarter says. "It would have the sensitivity to detect the warm backside of Dyson sphere collectors throughout the galaxy. If you were to build this thing and you didn't find anything, the limits that you set are pretty significant."

Alien visitors may have left evidence in a far less technological form—embedded in the DNA of life on Earth. Paul Davies, cosmologist at Arizona State University in Tempe, Arizona, proposes that interstellar visitors might have intentionally or unintentionally left bioengineered traces within terrestrial DNA. The self-replicating life forms would have passed down the biologically embedded messages, which still await our discovery.

Another solution to the Fermi paradox, says SETI's Jill Tarter, is that there is no paradox. "We've barely begun to look. If you take the nine-dimensional space that you might need to search through the electromagnetic signal, assuming that's the right thing to look for, and if you set the volume of that space equal to the volume of Earth's oceans, then so far we've looked at a small glass of water."

The Fermi paradox includes the idea that if there ever was a technological civilization anywhere in the Milky Way Galaxy at any time, they would have inevitably developed the ability for interstellar spaceflight. Most models for colonizing the galaxy show that the process would happen in a very short time—in terms of the lifetime of the galaxy—filling interstellar space with a cacophony of galactic communities. Fermi's paradox essentially declares that those civilizations are not present, and therefore there can't have ever been another technological race. The implication is that we are the first. But Tarter asserts that, "You can't say that. We have so poorly explored even our own Solar System that there could be *Battlestar Galactica* out there and we haven't noticed it. So when the Fermi paradox gets extended to the detection of signals, you get the same answer. We can't say there are no signals, because we've just started. You can't draw any conclusions on the basis of that. There isn't really a paradox; there's just a lack of observation. We're young in technology in a very old galaxy."

The time involved in interstellar communication may not be an issue for another kind of sentient being, those who are manufactured. Rather than biologically advanced species, the sole survivors of an ancient race may be its cognitive machines. Many science fiction stories (Asimov's *I, Robot*, the movie *Ex Machina*) project a time in the very near future when our computers become self-aware. Unlike us biological beings, thinking, aware computers would not fear the long, dark night of interstellar travel. If their nature incorporated curiosity, even these intelligent assemblages of hardware would explore the cosmos in search of other thinking beings.

As for our first contact with advanced beings via radio or optical means, we can only speculate on what the details will be. But what if SETI succeeds? Will we, as a sentient species, be changed forever philosophically? Will we want to go visiting? The next chapters will explore these questions.

8

Could We Visit Earths of Distant Suns?

The greatest problem that faces us when communicating with—or going to visit—alien civilizations is that of distance. With the centuries of time lag for round-trip messages, or the millennia involved in interstellar travel using conventional technology, the human race has been hard put to seriously consider face-time with distant races. In other cosmic settings, there may be locations whose physical layout has encouraged interstellar communication and travel, regions where radio communication between star systems takes months rather than years, and where travel to the nearest star might involve years rather than centuries. These locations reside in globular clusters, spheres of stars that tend to orbit larger galaxies (Fig. 8.1).

Stars in globular clusters are very ancient, quite stable, and long-lived. Because of the short distances between the member stars, and the fact that planetary systems there are very old, some researchers propose that the conditions are ripe for interstellar communication and travel. The stars in our neighborhood are few and far between. The close-by Alpha Centauri system, at 4.2 light-years away, would take generations to visit, even with technology that enables speeds far beyond our current capacity. But in the typical globular cluster Messier 30, several hundred thousand stars are packed into a ball just 90 light-years across. Typical distances between stars are less than a light-year, and many huddle as closely together as the width of our Solar System. In the core, star distances average about a fifth of a light-year. Some clusters are even denser, with a million stars swarming within a sphere just a little over 3 light-years across.

The stars in globular clusters tend to be tame by the Sun's standards, so radiation levels would be far less than the busy hub of our Milky Way Galaxy.

© Springer International Publishing Switzerland 2017   **193**
M. Carroll, *Earths of Distant Suns*,
DOI 10.1007/978-3-319-43964-8_8

Fig. 8.1 On its way out of the Solar System, a starship of the Enzmann design uses a gravity slingshot from Saturn and Iapetus to send it on a journey to a distant, Earthlike world. The 600-m interstellar craft is crowned by a 3-million-ton globe of frozen deuterium fuel. (Art © Michael Carroll)

The close distances would enable spacefaring civilizations to settle nearby systems, spreading through many planetary families in a cosmically short time period. By comparison, our stellar neighborhood is a dark, lonely place. And aside from travel, radio communication would be reasonable with round trip durations of a few months or years rather than the dozens or hundreds that it takes our communications to reach the stars. Rosanne DiStefano of the Harvard Center for Astrophysics posits that globular clusters have what she calls a sweet spot, a region where stars are so densely packed that their short distances enable easy interstellar communication, and even voyages. If civilizations have matured and moved out into their busy stellar neighborhoods, they may have established outposts. Those outposts ensure the survival of such beings, which means that globular clusters may play host to very ancient and developed civilizations. One of the major roadblocks to interstellar travel for us is the distance between the stars. But this barrier may not be enough in these busy stellar groups to dissuade sentient beings from communicating and traveling.

Interstellar Vacations

In our lonely region between the starry spiral arms of the Milky Way, the dream of travel to the stars is a challenge to our science and engineering—especially if one wants to go faster than the speed of light—but not a new one to science fiction. *Dr. Who* and the Time Lords had their Tardis, which linked space and time in very Einsteinian ways. *Star Wars* has its hyperdrive (which is never adequately explained), *Battlestar Galactica* its FTL (faster than light travel), and *Star Trek* its warp drive. *Star Trek's* concept comes closer to actual scientific concepts, describing a cosmic bubble that breaks through the fabric of space/time. Theoretical physicist Miguel Alcubierre put forth a similar concept in 1994. Instead of breaking the light-speed barrier, Alcubierre's "drive" would compress space in front of a ship and stretch it out behind. Alcubierre's work[1] is consistent with Einstein's field equations, but no one has come up with a way of actually doing it…yet. Frank Herbert's *Dune* series avoided an entire suite of technological stumbling blocks with its creepy creatures of the Spacing Guild, who fold space with the aid of an exotic spice. To some, this sounds like a return to college days.

Engineers today busy themselves with more conventional travel involving velocities slower than the speed of light. Less spice and more nuts-and-bolts. A host of propulsion systems are operating throughout our Solar System today. Ion drive and chemical propulsion are the primary drivers today, but

---

[1] "The Warp Drive: hyper-fast travel within general relativity" by Miguel Alcubierre. *arXiv:* gr-qc 0009013, September 5, 2000.

visionary aerospace engineers have also made headway in nuclear propulsion, laser or microwave thrust and solar sailing.

Even the modern trappings of interstellar travel came surprisingly early, as seen in this remarkable excerpt from *The Science Book of Space Travel* by Harold Leland Goodwin, written a full 3 years before the orbiting of the first artificial satellite in 1957:

> Such threads form the warp and woof of space travel, for the very nature of the subject embraces both sense and nonsense. For example, the velocity of light is accepted as a limiting factor…Our nearest star neighbor is Alpha Centauri, over four light-years away. Only a few stars are closer than a dozen light years. So simple arithmetic and common sense tell us that interstellar flight is beyond hope. Yet, with their breed's particular madness, men speculate seriously on voyages to other stars, and scientists and scholars discuss solemnly the possibilities of intelligent life on hypothetical planets of other solar systems. Only time will tell which thoughts are sense and which are nonsense.

In his introduction to *2001: A Space Odyssey*, author Arthur C. Clarke declares, "But the barriers of distance are crumbling; one day we shall meet our equals, or our masters, among the stars. Men have been slow to face this prospect; some still hope that it may never become a reality. Increasing numbers, however, are asking: 'Why have such meetings not occurred already, since we ourselves are about to venture into space?' Why not, indeed?"

Some aerospace teams and futurists are hard at work on the problem. It's a daunting one. For those of us not living in a globular cluster, near-light speeds will bring us to the nearest star in decades. Our fastest spacecraft to date travel at 1/10,000 the speed of light. At that rate, landfall at the Alpha Centauri system would occur in 100,000 years.

However, these painfully long travel times come at the hands of today's propulsion. On a mission like the New Horizons to Pluto, travel is much like throwing a rock. The vast majority of energy comes at the very beginning, with the booster tossing the spacecraft on its way. Small adjustments can be made along the way, and gravity assists from various planets can cut travel time by years. But what if we could constantly accelerate a large ship during its entire journey? A constant thrust changes everything.

For two decades, former space shuttle astronaut Franklin Chang Diaz has been out to do just that. He plans to change everything with the development of the Variable Specific Impulse Magnetoplasma Rocket, or VASIMR engine. VASIMR uses microwaves to heat propellant, turning it into a plasma.

VASIMR is an electric propulsion ion engine. Others have come before it, like the ion engines used on the Dawn and Deep Space 1 missions. Those electric engines used xenon as the propellant. Electricity heated the xenon to make a plasma. Then, using electric fields, the plasma shot out in an ion beam. Those ion engines had very low power levels. At roughly a kilowatt, they put out as much power as a hair dryer. Their advantage came not in strength, but in endurance over the long haul.

While working on the same principal, VASIMR is able to process far more power. The VASIMR engine's power can reach 200 kilowatts, the power of an SUV. If that power can be sustained, enormous speeds can be reached for interplanetary travel. A human mission to Mars could be trimmed from six or eight months (one way) to 6 weeks.

While VASIMR may be a game-changer for interplanetary travel, it still offers a snail's pace on the interstellar scale. Suppose we advance far beyond the VASIMR, increasing our speed to an acceleration that provides us with 1 g, or the equivalent of Earth's surface gravity. We would need to accelerate toward our destination only to the midpoint, at which time we would turn our ship around to begin deceleration. With this type of travel, a conventional six-month voyage to Mars would take about a day, and New Horizon's 9-year trip to Pluto would dwindle to less than two weeks.

Our new, tricked-out spacecraft, still traveling at only a fraction of the speed of light, would make travel in our Solar System easy. But the distances between the stars are vast. Our speed-trip to the nearest star would require a 4-year dent in our schedule.[2] That's a fairly respectable timetable, but well beyond current technology. To travel among Earths of distant suns, we need more speed. If we could bump our ship up to 99.5 % of the speed of light, our crew could reach the nearby star Vega in about 25 years. Due to relativistic effects, time would slow for those on board ship, making the journey roughly 1 year for them. For the round trip, they would need to carry only 2 years' worth of food and oxygen, and they would age only 2 years. But our starry-eyed tourists will face psychological challenges ahead. The bad news is that most of those they leave behind on Earth—family, friends, and coworkers—will have become deceased by the time the crew arrives at their destination. By the end of their round trip, Earth would be a very different world, 50 years advanced in technology and changed in culture. In the past 50 years, we have seen a progression from black and white broadcast TV to 3D streaming HDTV, from the Beatles to

---

[2] This would mean 749 days of travel in each direction. But this is an oversimplification. Because of relativistic effects, you cannot continue to accelerate at the same rate. As you approach the speed of light, your mass becomes greater and acceleration decreases.

Justin Bieber, from the 30-ton ENIAC computer (with 18,000 vacuum tubes) to the iPhone. Technological and social transitions move at an increasing pace. Imagine what changes will confront our future star-hoppers.

Time is only one problem with near-light-speed travel. The sustained power required for such journeys challenges the greatest engineering minds. The popular TV and movie series *Star Trek* has its starship *Enterprise* winging its way to all corners of the galaxy. But for a vehicle the size of the *Enterprise* (roughly the mass of an aircraft carrier), getting such a craft to only half the speed of light would take more than 2000 times the total annual energy use of Earth. Obviously, interstellar travelers will require new forms of propulsion. And there are dangers out there. At near-light speeds, even the benign particles of gas and dust floating through the cosmos could penetrate thick walls in much the same way that cosmic rays do.

As creatures living on a planet with a molten core, such particles don't worry us. Sloshing metals at Earth's heart generate magnetic fields. The biome of Earth is protected by this magnetosphere, a globe-enveloping bubble that shunts solar radiation away from the surface.[3] Since the 1960s, engineers have wondered if a similar field could be generated around a spacecraft to protect a crew. Recent studies in the UK, Portugal, Sweden, and the United States show promise. Engineers in the UK demonstrated that the technology could be made compact enough to shelter an inhabited spacecraft moving at breakneck speeds through interstellar space. Like a planet's magnetosphere, these local fields would generate a charge in the space around the craft, deflecting deadly particles away from its precious cargo. A recent study called SR2S hopes to develop several key enabling technologies to generate protective magnetic fields around a craft using superconducting magnets. Researchers at NASA's Johnson Space Center are experimenting with high temperature superconducting structures to generate protective fields. Engineers are working with a combined system of active and passive radiation shielding, with the ultimate goal of producing large, ultra-light, expandable super-conducting coils (Fig. 8.2).

Roadblock to Faster-Than-Light Travel

To date, we have no technology up to the challenge of faster-than-light travel. The problem is twofold: physical and physics. On the physical side, the roadblock is a simple matter of mass. Making a spacecraft go faster requires more

---

[3] Some of that radiation travels down the poles of our magnetosphere, causing the elegant glow of the aurora borealis and aurora australis.

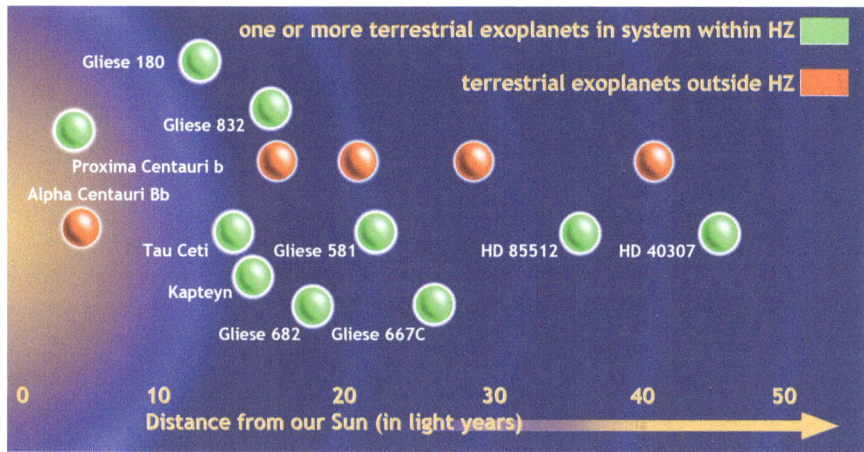

Fig. 8.2 How far would we have to travel to the nearest Earthlike terrestrials? Here, green spheres (labeled) represent terrestrial planets in their stars' habitable zones, while orange ones show terrestrials outside of their stars' HZ. Chart does not take into account size; some are smaller than Earth, while others are super-Earths. The closest terrestrials in their habitable zones are planets orbiting Tau Ceti (a Sun-similar but slightly smaller star) and Gliese 180 (an M-type red dwarf). (Art © Michael Carroll)

fuel, but that fuel adds mass, reducing the craft's capacity to accelerate. As for physics, as a spacecraft gains momentum and approaches the speed of light, rules of relativity demonstrate that its mass continues to increase as well. As an object reaches near-light-speed, its mass becomes infinite.

There may be ways around the first problem. The writing staff of *Star Trek* gave a nod to one: the use of matter/antimatter mixing. The spectacular efficiency of nuclear thrust is overshadowed by this scheme. Nuclear fusion converts roughly 1 % of its atomic fuel's mass into energy. Theoretically, the annihilation of matter/antimatter converts 100 % of its mass into pure energy. This futuristic propulsion could enable ships to reach over 90 % the speed of light. At that speed, time slows substantially for the crew due to relativistic effects (the faster you go, the more slowly time seems to pass). Because of this effect, the trip to nearby stars would compress for the crew, bringing nearby star systems within easy reach. The problem is that we have no natural sources or storehouses of antimatter. Tons of the strange material would have to be manufactured and then carefully stored, both of which are beyond our current technology.

However, the mass/velocity ratio is a challenge that appears to have no conventional solution. It seems that for the foreseeable future, engineers will have to settle for speeds more leisurely than that of light. Despite some seemingly insurmountable challenges to even conventional travel, visionary designers have been proposing serious interstellar projects since the 1960s. Engineers first focused on unpiloted, robotic probes.

Project Orion

Going the interstellar distance called for new ideas in propulsion. The power of conventional chemical and ion drive engines paled when compared to that of nuclear power. Nuclear power is on the order of a million times as potent as chemical energy. Controlling that energy was the challenge, but designers began to craft blueprints in the 1960s, incorporating the power of nuclear explosions. Some studies went beyond the drawing board to the stage of field testing. One such project would have easily lofted a payload the size of the starship *Enterprise*. It was called Orion. It was an apt name. Orion was the man in the winter sky, the great hunter, the adventurous explorer.

The genesis of Orion rested in the idea of exploiting peaceful uses of atomic power after World War II. At General Atomic, Ted Taylor—who helped to design the first atomic bomb—led a team of fifty scientists to build a nuclear-pulse interplanetary vehicle. The impulse propulsion would pulverize a small ship, but a huge one, built with the same techniques and materials used in conventional shipyards, could carry a massive structure into space and on to the planets.

In a redefinition of the phrase "plows into plowshares," the Orion project would conscript the energy of thousands of 1-kt nuclear detonations. To shield the payload from the jolt and radiation of the blasts, a "pusher plate" at the base of the craft would take most of the punishment. Detonations would occur a few tens of meters behind the ship. Technical studies at the time projected that with 1000 small explosions, the ship could make it to orbital velocity. Under a steady stream of pulses, the pogo-stick craft could pass the Moon in just 5½ hours, and out to Jupiter and its moons within a year. The numbers were so promising that the team developed the motto "Saturn by 70."

To put Orion's scale in perspective, it took the world's spacefaring nations 45 launches to loft all segments of the International Space Station into orbit. Orion could have lofted two of the 500-ton stations in one launch. Actual field tests carried out on Point Loma in San Diego demonstrated the feasibility of the exotic nuclear-pulse propulsion system. With this adaptation of nuclear weaponry into power for exploration, astronomer legend Carl Sagan said, "The Orion starship is the best use of nuclear weapons I can imagine."

Orion's baseline blueprints called for three main components. At the base stood the nuclear jettison propulsion module, which included the pusher plate. Above it, the stacks of magazines containing the nuclear pulse units were arrayed in rows for ejection into the propulsion module. Atop the entire vehicle rested the payload stack.

Visionaries put several versions on their drawing boards. The original concept would have been launched from the ground, rising on an incandescent—and radioactive—column of plasma. The estimated power was so great that launch weight was no limiting factor. In fact, engineers utilized off-the-shelf submarine and battleship-hardened parts for the craft.

A second, smaller version of Orion could have been lobbed into orbit on a Saturn V or similar booster. Once in orbit, the Orion ship would begin its chain of explosions to send it on its way. Other scenarios made use of rings of solid boosters around the body of the vehicle.

Orion was originally intended for interplanetary travel. Its creators envisioned a nest of dozens of probes aboard ship, each being deployed to a different planet or moon by an advanced artificially intelligent computer. But the ship could be adapted to interstellar travel as well. Projections of such a scenario indicate that an interstellar ship accelerated by the rapid-fire explosions of roughly a million hydrogen bombs could reach Alpha Centauri in a little over a century.

The biggest challenge for Orion was to fabricate thousands of cheap, portable nuclear bombs for the project. Orion would need to deploy a dozen every minute in some scenarios. Although the system seems intimidating, it is significant to note that in the 1960s, researchers considered the process solved from an engineering standpoint. It must have shown promise—the work is still classified.

Another concern voiced by critics was the electromagnetic pulse (EMP) associated with nuclear blasts. EMPs can wreak havoc with power grids, cell phones and the Internet. But Orion's charges are a fraction of the power that causes these types of problems.

Despite the promise Orion showed, its design had an intrinsic drawback—radiation fallout. The radiation from an Orion launch would be equivalent to one surface test of a moderate nuclear bomb, resulting in one to ten deaths globally from the increased worldwide radiation. That was one to ten too many.

The world's nuclear test ban treaty put an end to practical testing of Orion in 1963. Another consideration was security. A small, portable nuclear device is just the technology that terrorist groups would like to get their hands on (Fig. 8.3).

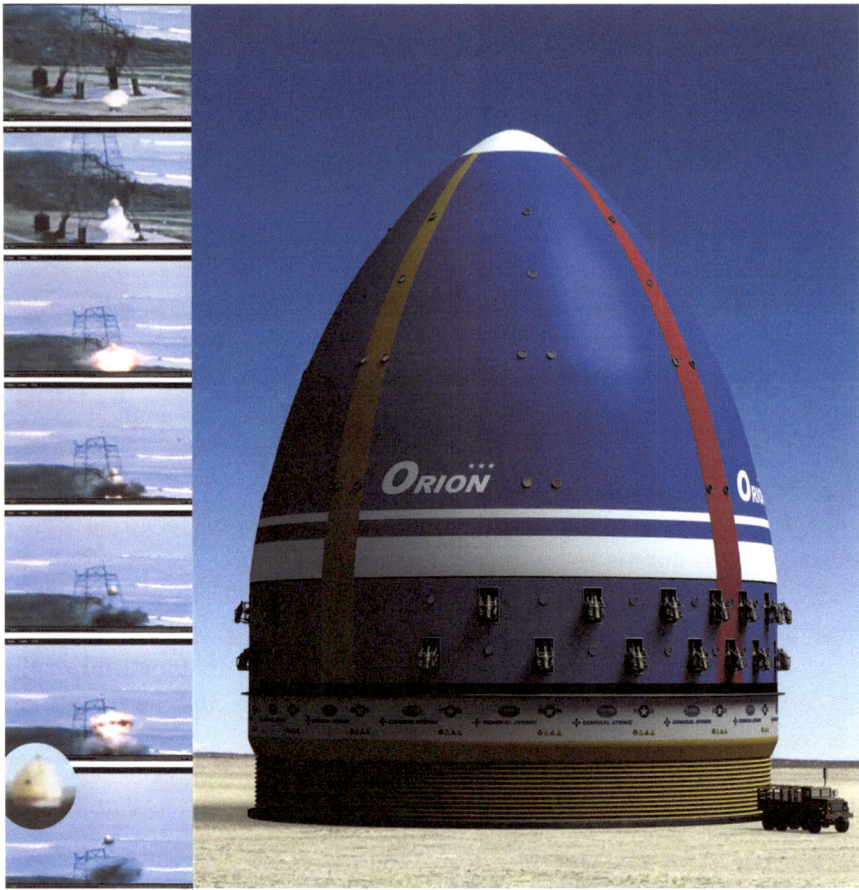

Fig. 8.3 *Left:* Frames from 8-mm vintage footage of an actual Orion model test launch using C4 explosive. Successive blasts propelled the scale model, which remained intact in all trials. (Image courtesy of the National Archives.). *Right:* Artist Nick Stevens envisions the Orion interplanetary craft ready for take-off. (Image © Nick Stevens. Used with permission)

Project Daedalus: To the Stars

While the world's political arena may not have been ready for nuclear pulse propulsion, the engineering realm realized its potential. Studies continued quietly. From 1973 to 1977, the British Interplanetary Society (BIS) embarked upon a study of an Orion-based robotic ship. Their technical study chose, as its target, Barnard's Star. Barnard's Star is an M-type red dwarf star 6 light-years from Earth. Aside from the Alpha Centauri system, it is the closest sun to our own. In the 1970s, astronomers did not yet know whether Barnard's had a planetary system, but several studies suggested that it did.[4] As a single star, and for the purposes of the study, it would do nicely.

The BIS study indicated that nuclear pulse propulsion could bring a spacecraft up to 12% of the speed of light in just 4 years. At that speed, the craft could arrive in the Barnard system within less than half a century. As it approached, it would map the entire system of planets and moons and determine a strategy for deploying up to twenty planetary probes. The probes would need to depart several years before the craft arrived in the environs of Barnard's Star, so that each could reach its individual destination. An erosion shield would crown the front of each probe, with ion engines at the rear.

Instruments linked to powerful telescopes would instigate remote sensing as each probe carried out a dedicated encounter with targeted planets and moons. Sub-probes could be dispatched to perform close studies and landings. Data from all the encounters would be fed to the mother ship for study by the ship's artificial intelligence (AI), with eventual transmission to Earth. Because of the enormous lag time in light-speed radio communication—a one-way message would take 6 years—the ship's AI would determine how best to study the various worlds it encountered.

The so-called Daedalus study constituted a first approximation of interstellar missions. The project represented a formidable task, its vehicle dwarfing the Saturn V booster, largest to ever fly. Weighing in at 54,000 tons (20 times that of the Saturn V at launch), Daedalus could carry a payload of 500 tons to Barnard's Star system (Fig. 8.4).

The baseline study used pellets of deuterium and helium-3 for propulsion. Unlike Orion's pusher plate arrangement, the blasts would take place in a chamber surrounded by a magnetic field. The pellets would be ignited in the center of the chamber by electron beams surrounding it, facing inward, and shunted out the back for thrust. The rate of ignition would be 250 pellets each second. Because helium-3 is such a rare isotope, its actual mining[5] is one limiting factor of the design.

The Daedalus ship's structure would be dominated by a ring of giant tanks around its midsection. Each would hold millions of frozen deuterium/helium 3 fuel pellets. As fuel was exhausted, empty tanks could be jettisoned to rid Daedalus of the dead weight. In the hollow at center of the tank ring would be the craft's service facilities, including power supply, chemical propellant for attitude correction, and a bevy of "wardens," robotic maintenance droids that

---

[4] Research by ground-based and Hubble Space Telescope observatories has determined it unlikely that Barnard's system retains planets of near-Jupiter size, and even terrestrials in the HZ are unlikely.

[5] Helium-3 is available on the surface of the Moon and in the atmospheres of the gas and ice giants.

Fig. 8.4 Project Daedalus, an interstellar outgrowth of project Orion, depicted by Nick Stevens. (Image © Nick Stevens, used with permission)

tended to the planetary probes, checked for leaks or structural failures, and carried out repairs. An erosion shield would cap the front of the interstellar ship. With its great speed, Daedalus would be subject to damage by even the minutest of particles. Behind it would be the payload bay, loaded with the disposable probes, instrument platforms for remote studies, and the ship's main computer (Fig. 8.5).

Project Icarus

Project Daedalus demonstrated proof of concept for many technologies that were deemed current or within plausible extrapolations of existing technologies. Daedalus provided a reasonable projection of a scenario to launch an

Fig. 8.5 Massive ships compared (*left to right*): Daedalus, Orion and Saturn V, largest booster to date. (Image © Nick Stevens, used with permission)

interstellar mission that could reach another planetary system "on the timescales of a normal human lifetime." One goal of the British Interplanetary Society today is to advance the kinds of technologies that could lead to interstellar travel. To that end, the society launched a global effort called Project Icarus on September 30, 2009.

Icarus was a 5-year theoretical engineering study for the development of an interstellar craft. According to the BIS announcement, its purpose was four-fold: (1) to motivate a new generation of scientists in designing space missions that could explore beyond our Solar System; (2) to generate greater interest in the real-term prospects for interstellar precursor missions that were based on credible science; (3) to design a credible interstellar probe that was a concept design for a potential mission in the coming centuries so as to allow a direct technology comparison with Daedalus, and (4) to provide an assessment of the maturity of fusion-based space propulsion for future precursor missions.

Icarus was and remains a key project of the non-profit Tao Zero Foundation. In concert with BIS, Tao Zero's collection of scientists and visionaries are dedicated today to the incremental development of interstellar spaceflight. Rather than wait for all the engineering pieces to fall into place, Tao Zero has embarked upon the study of small probes that will utilize and demonstrate interstellar propulsion technologies. The team believes that much can be learned by sending spacecraft to intermediate targets much closer to home that offer challenges.

Another mission that has gotten Tao Zero support is called the FOCAL mission. FOCAL makes use of one of the techniques common to exoplanet-hunting, that of gravitational lensing. FOCAL would utilize the Sun's own gravitational lensing power, at a distance of some 550 AU out, to image not only other stars but also their planets and moons. Italian physicist Claudio Maccone has been making the case for FOCAL for decades. At a recent Princeton conference called "New Trends in Astrodynamics and Applications," Maccone told attendees that, "As each civilization becomes more knowledge-able, they will recognize, as we now have recognized, that each civilization has been given a single great gift: a lens of such power that no reasonable technology could ever duplicate or surpass its power. This lens is the civilization's star. In our case, our Sun." Tau Zero would like to confirm the physics and use of Maccone's gravitational lensing using the Sun beyond 550 AU. Such missions would be able to resolve nearby star systems in great detail, affording mission planners with orbital, physical, and even geological and atmospheric charac-teristics of systems under consideration for an interstellar probe. The beauty of a FOCAL mission is that a space telescope need not be stationed exactly at the 550 AU distance. As the observatory makes its way farther out, the effect of the gravity lens becomes even more powerful.

Non-Nuclear Options for Star Travel

If exotic nuclear pellets are not your style, starship designers are studying other choices. Another type of interstellar propulsion is called an interstellar Bussard ramjet. The concept, first popularized by physicist Robert Bussard, entails a giant scoop at the front of the spacecraft used gather interstellar hydrogen for use as fusion fuel. In this way, the ship would not need to carry the weight of fuel storage, but would run fast and light, gathering its fuel along the way. Because the hydrogen between the stars is so rarified, the scoop would need to be titanic. Rather than building such a large physical structure, magnetic fields generated by the ship would corral hydrogen from a wide spread. Electromagnetic fields could extend for hundreds or thou-sands of kilometers out, harvesting and compressing the hydrogen. The ship's high speed would pack the hydrogen as it funnels into ever-smaller areas, finally compressing it to the point where it triggers thermonuclear fusion. The exploding hydrogen would move through the ship's center, venting out the back as rocket propulsion.

The Bussard ramjet has some physics problems. Some studies show that no system could accrue enough hydrogen fuel to offset the pressure of the solar wind, or that the hydrogen between stars is too thin to create efficient thrust. But work continues. The problems with Bussard's original concept led to the Ram Augmented Interstellar Rocket (RAIR). RAIR carries its own nuclear fuel supply that generates thrust. But its nuclear reactions are supplanted by interstellar hydrogen, which it scoops up along the way. RAIR is a three-part system of a fusion reactor, a Bussard-style scoop, and a plasma engine. The reactor supplies the electrical power to ignite the incoming hydrogen (Fig. 8.6).

Interstellar ramjets could get help from launches ahead of time. Several years before the departure of the interstellar craft, Earth could launch hydrogen fuel "caches" along the projected journey. As the ship gained speed, it would overtake these more slowly moving payloads. Each of the caches would be disbursed into the space around them shortly before the mother craft reached the region, seeding space with more fuel for the oncoming traveler.

Other proposals have been put forward to get us to our stellar destinations. In what Freeman Dyson called "the reincarnation of Orion," NASA and the U. S. Air Force studied spacecraft powered by microwaves. Powerful beams of energy in the form of lasers or microwaves—sent from Earth or nearby— would fall on a large sail-like structure to provide thrust during launch and in interplanetary and interstellar flight. In their study, 4½ minutes of microwave blasts from an orbiting station would provide 3 g's of acceleration.

During ascent through the atmosphere, the laser explodes air beneath the craft. In tests in 2000, engineers at the High Energy Laser Systems Test

Fig. 8.6 A Bussard ramjet pulls interstellar medium into its scoop for fuel. (Image © Nick Stevens, used with permission)

Facility proved the concept. A specially designed, dish-shaped reflective surface focused the laser into a ring at the center, propelling the craft with the exploding atmosphere below it. On October 2, 2002, lasers lofted the tiny test craft (just 12 cm in diameter and 51 g), blasting it to an altitude of 71 m. The flight lasted 12.7 s. Designers estimate that by increasing the laser's power to 100 kilowatts, a similar craft could be launched to an altitude of at least 30 km. Eventually, engineers plan to send a 1-kg microsatellite into orbit, powered by a 1 megawatt laser.

Once in space, the propulsion takes on the form of a powerful, focused solar sail. Solar sails, which are launched conventionally and deployed in the vacuum of space, are completely passive, using the pressure of sunlight to move as a sailboat uses wind to traverse a lake. Several successful solar sail missions have been carried out to date. Japan's space agency, JAXA, launched the world's first interplanetary solar sail mission, called Ikaros, in 2010. The 46-foot/14-m-wide kite-shaped sail is powered by solar cells embedded in the sail's fabric. Using a set of liquid crystal panels to change the surface reflectivity, the craft is able to change its orientation. Current flowing through the LCD panels increases reflection for Ikaros to move forward, while turning the flow off reduces the pressure of the sunlight. The Planetary Society's privately funded LightSail 1 was a technology demonstrator. After launch on May 20, 2015, and complications that followed, the sail was fully deployed on June 7. Images from orbit showed the craft had successfully erected its structure and was maneuvering in space. The society plans a follow-on, LightSail 2, now scheduled for a SpaceX launch in March of 2017. For its part, NASA plans to launch the world's largest solar sail, called Sunjammer. The mission is named for a solar-sailing ship in an Arthur C. Clarke short story about a solar sail regatta from Earth to the Moon. Sunjammer will span 38 m on a side. Mission goals include a technology demonstration, but the craft will also be used to monitor space weather.

For interstellar travel, a craft could be fashioned with vast gossamer sails to catch the power projected from close to home. A lightweight interstellar craft 50 km across would set sail riding a wave of laser light or microwaves generated by power satellites near the Sun. Studies have determined that a power satellite 1.2 km^2 across could generate 10 billion watts of power. The lasers would be stationed in near-Earth space, powered by solar energy collected by large reflectors. The reflectors would direct their beams toward the spacecraft, filling its sails with the pressure of light (Fig. 8.7).

In the case of an interstellar ship, lasers would be more efficient and powerful if they were in orbit close to the Sun or circling Mercury. The laser light produced near the Sun by a constellation of satellites could be woven into a single coherent beam and sent out to a focusing lens stationed in the neighborhood of Saturn or Uranus. From there, the beam could be directed at the interstellar craft for vast distances. Some estimates put the flight time at a millennium, so patience would be in order for any flight engineers.

Interstellar tall ships are not needed to make our first visits of distant Earths. Our initial surveys could be carried out by tiny craft weighing just a few grams. Flights of miniature interstellar probes are the brainchild of physicist and writer Dr. Robert L. Forward, and were further developed by NASA Glenn Research Center's Dr. Geoffrey Landis (also an award-winning author). Forward assumed that even moderate microwaves would be sufficient to propel a small craft that he called Starwisp. He determined that a superconducting metal mesh with a sail mass of 16 g and a payload mass of 4 g would do the trick. The total mass of his Starwisp probe would be a humble 20 g. Landis called for a similar sail design fabricated of carbon wires with a larger sail mass of 1000 g, a payload mass of 80 g, and a diameter of 100 m.

At the urging of its interstellar microwave beam, the probe's acceleration would reach 24 m/s^2. Its power would flow through a microwave lens 560 km in diameter, transmitting 56 GW of power. Eventually, the Starwisp would accelerate to 10 % of the speed of light.

Microwave energy has an advantage. It can be made and transmitted at extremely high efficiencies, although it is difficult to make narrow beams that extend over long distances. Because of its weakness over distance, the star probe would have to accelerate quickly to reach the high velocities needed for interstellar travel, gaining momentum before it got too far from its microwave transmitter. The accelerations of a craft such as Starwisp are too high for a human crew to survive, so microwave-thrusted probes may only be robotic. The Starwisp's sail is a wire mesh containing microcircuits at the intersection of all the wires.

The microwaves that power the starship have a wavelength that is larger than the openings in the wire mesh. This means that the lightweight wire netting acts like a solid metal sail in the microwave's beam. The amount of push is miniscule, but if the sail is light and the power in the microwave beam is high, the probe can accelerate to nearly the velocity of light while the starship is still near the transmitting lens in the outer Solar System.

As Starwisp reaches its target star, the microwave transmitter back in the Solar System is turned on again, pouring microwave energy into the star system. The wire netting now serves a new purpose: it gathers the microwaves for energy. Tiny microcircuits and miniature instruments on Starwisp, energized after a long flight through the darkness of interstellar space, carry out their reconnaissance of the planets, moons, asteroids and comets in the system.

Fig. 8.7 A sail ship moves under the power of microwaves beamed from its home world. (Art © Michael Carroll)

It then uses its wire mesh as an antenna to beam back our first close-up views of worlds around a distant star.

Other high-tech advances are opening the door to light-powered star travel at significant fractions of light speed. The goal of a new project, called Breakthrough Starshot, is to demonstrate proof of concept for ultra-fast light-driven nanocrafts.

Breakthrough Starshot has been given financial support by popular cosmologist Steven Hawking. He has combined forces with Yuri Milner once again, to the tune of $110 million. At the announcement of the project, Hawking said, "With light beams, light sails and the lightest spacecraft ever built, we can launch a mission to Alpha Centauri within a generation...Today, we commit to this next great leap into the cosmos....our nature is to fly."

The Starshot spacecraft is tiny, comprised of a wafer-size chip attached to a diaphanous, reflective sail. The plan is to launch a pair aboard a single booster. Once in orbit, the microspacecraft would be propelled by laser light beamed from a high-altitude station near Earth's equator. Models project that the little craft could reach 20 % the speed of light, arriving at the Alpha Centauri star system in just two decades.

With substantial progress, Starshot could become the first interstellar mission, departing for Alpha Centauri within the next generation. En route, the twin spacecraft could contribute to other sciences, including Solar System surveys in the outer planets, the search for Earth-crossing asteroids, and the mapping of the Sun's heliosphere structure and boundary.

A host of difficult technological and design challenges face engineers on the project. The Starshot team has listed them in an open access forum in hopes of generating helpful input from the aerospace community. Challenges include communication (tracking Earth with the transmitter, sending images with laser through the sail itself, enabling Earth lasers to receive images, etc.), miniaturized components (including cameras, photon thrusters, and processors), fabricating a light sail that will retain its structural integrity under the thrust of the laser, battery technology that will be reliable over a 20–30 year mission, keeping the sail stable within the beam, and the effects of interstellar dust and cosmic rays at the speeds called for.

Arrival

But once it gets to its destination, what would our robotic Starshot or Daedalus do? Australian physicist Ronald Bracewell gave the issue some thought. Bracewell was the inventor who devised the system now used by observatories to digitally block out the light of a star to search for planets. He envisioned an automated craft that could search out technologically advanced cultures in the cosmos.

As in the BIS study, Bracewell proposed that an interstellar probe could be built with advanced artificial intelligence (AI). The craft would be tasked with not only searching for other races in the galaxy but also with contacting them and dispensing information from our own. Naturally, we might discover the existence of distant advanced races by discovering their Bracewell probes as well.

Bracewell conceived of entire flotillas of probes sent out to the nearest stars using powerful interstellar propulsion. Traveling at a substantial fraction of the speed of light, the probes would eventually settle into orbit around stars containing biologically promising planets within habitable zones. Each star traveler would arrive at an orbit within the HZ, circularize its path, and begin to recharge itself with the light of its target star. It would then bring systems to bear in a search for artificial radio transmissions or other radio leakage from intelligent inhabitants. If the probe discovered any, it would record them, send data back toward Earth, locate the source of the alien transmissions, and beam the contents back to the source planet, unaltered. With its proximity to the planet, the probe's signals would be far more powerful than those coming from a star light-years away. Its echoed transmission would get the attention of its audience and announce the probe's location, hopefully triggering a dialog of some kind. If the strategy of mirroring the planet's transmissions did not work, the probe would then send its own message using a wider range of frequencies.

In a sense, a Bracewell probe would act as an ambassador from Earth. From its nearby perch, the AI could carry out a conversation in real time, circumventing the frustration of messages sent from light-years away. Such a vehicle could serve as a library, a storehouse of humanity's combined knowledge through the ages. With the ship's artificial intelligence playing host, the encountered race would be treated to the biological history of planet Earth. The origins of our civilization, its aspirations, monuments, technological and architectural accomplishments, would all play out in a rich panoply. If the craft were large enough, it could even carry artifacts and gifts from its home world. One drawback is that the conversation would be limited by the extent of the ship's knowledge base and artificial intelligence at its launch.

If, at our comparatively primitive stage of space exploration, we humans are thinking about wandering Bracewell probes, it is possible that advanced races have also done so, and have deployed such devices into our very own Solar System. Several telescopic searches have been undertaken, especially at Earth's Lagrangian points, in search of such robotic explorers. Lagrangian regions are gravitationally stable "parking lots" in space where any craft will remain at a fixed location compared to Earth as it moves through space. No alien probes have yet been found.

The concept of advanced alien civilizations placing monitors near worlds where intelligent life is likely to arise is called the sentinel hypothesis.[6] Robot sentinels may be charting the progress of the human species as we speak. Automated sentinels could dispatch smaller scout ships to monitor planetary systems around stars, keeping an eye on planets with promising environments. The larger sentinels, alerted to the presence of civilization by the scout ships, could be programmed to make contact once a civilization advances to a prescribed level. Such milestone advances might include certain levels of radio transmissions or the development of interplanetary flight.

---

[6] Not to be confused with the sentinel theory of REM sleep.

A variation on Bracewell or sentinel probes is the von Neumann probe. Mathematician and physicist John von Neumann was a pioneer in the field of self-replication. He applied his theories to robotic space exploration, coming up with the concept of the von Neumann self-replicating probe. Like the Bracewell probe, von Neumann's brainchild would travel interstellar distances to contact neighboring cultures. He visualized a craft consisting of a propulsion system, a sophisticated AI, and a "universal von Neumann replicator" to create copies of itself. After its arrival at the target star system, the craft would search for raw materials from which to "print" its descendants. These materials could be found in a star's equivalent to our asteroid or Kuiper Belt. After mining for local construction supplies, the craft would make several copies of itself (including propulsion system and robotic brain, onto which it would load the entire contents of its AI), dispatching them to further systems. Then, the ship would set to work exploring the planetary system around it. It would beam science results back to Earth for analysis, and seeking out intelligent races with which to exchange information. Meanwhile, the probe's offspring would be on their way to repeating the process in nearby, unexplored star systems.

Von Neumann probes could also be arks, carrying artificial biomes within them to plant on promising exo-Earths. Egg cells from the planet or origin—or genes built by the probe's own memory—could be implanted in the environment of the new Earth. As with Bracewell probes, the discovery of alien versions of von Neumann self-replicators would be evidence of spacefaring civilizations intent on making contact.

Interstellar Exploration's Human Component

As Earth's engineers and visionaries pondered the robotic exploration of other star systems, their attention naturally turned to the next logical step, human interstellar travel. During the days of Orion, the project employed hundreds of engineers, scientists and even farmers. The dream was as big as the 1000-ton ship—to carry a self-sufficient colony to the planets. The settlement contained within this uber-Orion would house between 200 and 400 inhabitants, a colony to explore the Solar System. With its constant acceleration, it was even possible to have open lakes, farm fields, restaurants and shops. With a ship of 1000 tons, designers could afford to be extravagant.

As is often the case, the military took an interest in Orion. A U. S. Air Force version fleshed out the details of Orion's propulsion module. A small cannon would fire nuclear pulse charges through a port in the pusher plate. With every jettisoned charge, the cannon's opening would seal before detonation.

Exactly 138 half-strength charges would begin a more gradual acceleration. Once the craft was out of the atmosphere and up to speed, full nuclear pulse units would replace them, exploding with full force. A magazine stack would hold 60 nuclear charges, and there would be 60 magazines in a layer, holding a total of 360 bombs. The air force design would hold up to ten layers. A full complement of ten layers would contain 3600 pulse unit explosives.

Above the dangerous nuclear pulse units would sit the payload, with three modules. Plans called for these to house a powered flight station, personal section, and a 12-m spine structure. The spine would support the magazine stack, which led to the propulsion module. It also could carry any mission-specific cargo, spare parts and a repair station. Above the spine would rest the personnel compartment, with life support, crew quarters, workshops and, in the case of a research vehicle version, laboratories and science equipment. Above the crew compartment, a "Powered Flight Station" would house a radiation storm shelter where a control room would guide the ship during acceleration. The shelter would separate crew from nuclear blast radiation, but could also be used in the event of a solar event (like a radiation-rich flare or coronal discharge). The shelter/control room also could serve as a detachable lifeboat. It would have a separate propulsion system, and life support for the 8-member crew to last for 90 days.

Another design for human-piloted interstellar flight is called the Enzmann starship (Fig. 8.1). This concept was proposed in 1964 by physicist and polymath Dr. Robert Enzmann. Enzmann envisaged a vehicle whose main feature is a 3-million-ton ball of frozen deuterium. The deuterium would fuel nuclear pulse propulsion engines similar to those of Project Orion. Behind the metal-covered sphere of ice, a cylindrical section would extend aft. This section would contain the crew quarters and modular living areas, held within three identical 90-m-long cylindrical modules. The Enzmann ship could support a crew of 200, but would have room to grow.

Everything about the Enzmann starship would be larger than life. The 600-m craft would be longer than the 449-m high Empire State Building. Its seamless metallic fuel tank would be tricky to construct. Enzmann proposed the inflation of a giant plastic balloon in orbit, coated with some type of durable metal. The spacecraft would need to be assembled in Earth orbit. Enzmann suggested that precursor missions be sent out to the target stars, along with all the near-Earth observations that would be possible.

Like Orion, shock absorbers on an armored "pusher plate" would be mounted to the base of the spacecraft. A continuous stream of small (~5–15 kt) detonations would provide thrust.

Human passengers create complications on an interstellar mission. They must eat and breathe and live. Near-relativistic speeds slow time, aiding in the problem of the time element, but to voyage to another star will take generations. If we are to use conventional engineering, even of the kind we can reasonably dream of with projected technologies, interstellar travel is a long-term effort. We may need to be content with sending a ship on a journey of hundreds of years. Such a ship would arrive with the descendants of its creators. Hence, this type of vessel is called a generation ship.

Generation Ships

Just as advanced civilizations might build cylinders or torus-shaped structures to house colonies within their interiors, interstellar shipbuilders might construct vast rotating ships to carry populations of humans and wildlife to other worlds. Like the O'Neill colonies, the massive structures would spin so that their contents would be forced onto the walls. The ship's center would always seem to be "up," while the inner walls would serve as the ground. These futuristic "Noah's Arks" have been called generation ships, because they would require many generations of travel to reach their destinations (Fig. 8.8).

Visionary author Kim Stanley Robinson describes the travails and triumphs of such a journey in his book *Aurora*. Robinson's ship is powered by a series of explosions, much like the Orion and Daedalus designs, augmented by microwave propulsion, with its arrival back home braked by a magnetic "parachute." His star voyagers endure a journey of nearly two centuries—seven human generations—during which the ship's 2000 inhabitants, quantum computer and robots must tend to all the ship's systems and various environments. Those environments include 24 different habitats with ecosystems named Nova Scotia, Sonora, Amazonia, Olympia, and others. The author even envisions a stone-age culture in an Arctic habitat. Each section has a population of wild creatures, with a careful balance between predators and prey. Even the lakes have marine life, and docks for sailboats. The novel's narrator is the ship itself. Here, it describes the motivations for building the interstellar craft:

> Expression of their burgeoning confidence in their ability to live off Earth, and to construct arks that were closed biological life support systems…That a starship could be built, that it could be propelled by laser beams, that humanity could reach the stars; this idea appeared to have been an intoxicant…They did not care as much about their descendants as they did about their ideas, their enthusiasms.

Fig. 8.8 The curving landscape of a generation ship would become the norm to generations of interstellar voyagers. Artist Bryan Versteeg envisions a golf course and ponds (*above*) and a sports field (*below*). Recreation and verdant environments will contribute to the health of a generation ship's passengers. (Paintings © Bryan Versteeg; http://www.spacehabs.com. Used with permission)

Robinson goes on to explore what the descendants of those early travelers feel about their fate, and about the adventure of making a home on an Earth of a distant sun.

For a ship traveling at a substantial fraction of the speed of light, the far end of the journey is a problem. After decades or centuries of acceleration, how does one stop? One obvious strategy is to turn the ship around at the halfway point in the journey and decelerate for the duration, but this makes the trip long, and it requires a lot of fuel. Plus, this strategy only works for ships that are self-propelled. But what about those being sent by microwaves or lasers? Theoretical physicists have stumbled upon a solution using the natural resources at hand—the magnetic fields of stars and planets.

A magnetic sail, or parachute, is essentially a loop of superconducting tether. When current passes through the tether, it creates drag against the background magnetic fields intrinsic to the approaching star. Loops of conducting wire tend to force themselves into a circle because of their self-generated fields, so the deployment of such a structure would be easy, as it would extend itself into the needed shape.

Acting like a sea anchor, the magnetic sail benefits from a force of nature called the Lorentz force. The Lorentz force emanates from electromagnetic fields. The force can be used to nudge a spacecraft in the right direction, if it has the right hardware. A starship could, theoretically, extend a hundred or so metal cables, each loop stretching out for hundreds or thousands of km. The crew would then charge up these cables with 800,000 volts. The current's interaction with the magnetic fields will cause the ship to make a huge, gradual pirouette around the target star so that it is eventually approaching from the far side. Years before the craft is in position, lasers or microwave beams can be turned on again, timed to reach the star system at just the right time to slow the ship as it approaches the star from behind (now traveling in the direction of Earth). The magnetic sail concept enables a ship to divest itself of roughly half of its velocity—even if that velocity is nearly relativistic—without the use of propellant.

Despite the lure of finding biomes on distant Earths, when humans seek a place to settle, active indigenous biology may be bad news. The first reason is simple: human settlers cannot afford to risk infection from an alien source. The consequences could be devastating. Secondly, no matter how careful they are, the people of Earth will carry a wide spectrum of microbes with them, and these could destroy any in situ biome we visit. In fact, some astrobiologists advocate a position of extreme care for the planet Mars, suggesting that if life forms are found there, all human exploration endeavors should cease immediately.

Why Go?

In the final analysis, it may be that interstellar travel is healthy for us, culturally, scientifically and technologically. The Apollo project endowed human society with technological advances, but it also helped a populace engrossed in conflict and tension to look beyond Earth, to aspire to something higher. In 1968, during a year that had witnessed film of a war-ravaged Vietnam, political unrest in Europe, and unprecedented social unrest in the United States, a remarkable image came back to the people of Earth. *Apollo 8* sent the first human crew around the Moon, and as the crew looked back and saw Earth in its entirety for the first time in history, the people across our planet were captivated. After the flight, astronaut Frank Borman received a telegram from a casual observer, a person he had never met, that simply read, "Thank you, *Apollo 8*. You saved 1968." Such is the power of exploration and the contemplation it brings.

"It's more than curiosity," adds Seth Shostak, "because societies that don't explore usually get swallowed up by those who do…Exploration has always been a good thing for any society to do."

9

First Contact: What Will It Mean?

Perhaps one day, we will make contact. Our patiently listening SETI engineers might receive an unmistakably artificial signal, or detect a vast structure brimming with living, broadcasting beings. Perhaps one of our robotic ships will make first contact with another race, or one of our generation ships will come face to face with a sentient, alien culture. What will it all mean to us, scientifically, culturally and philosophically? And why do so many people care?

SETI's Seth Shostak proposes that our search for interstellar companionship may have its roots in our very genetic makeup. "I think we're hard wired to be interested in other critters that are sort of like us. It could have a component of not wishing to be in all of this alone. That's consistent with the data; it could be that the universe has no bounds. That's a lot of universe and to think that this is the only place where anything interesting is happening could be, in some ways, disturbing."

In their famous 1959 *Nature* paper, authors Morrison and Cocconi pointed out, "Few will deny the profound importance, practical and philosophical, which the detection of interstellar communications would have. We therefore feel that a discriminating search for signals deserves a considerable effort."

However, some have their doubts about the great search for cosmic companionship. Doubters declare that SETI is too expensive in a world wracked with poverty and war. But to put the cost of SETI into perspective, in 2015 the combined efforts of the search for extraterrestrial intelligence cost less than two Block IV Tomahawk Cruise Missiles.[1]

---

[1] This is $1.4 million each in 2015 USD, according to congressional records; $550,000 according to the U. S. Navy website. SETI 2015 budgets, including all private funding, reached just under $2.5 million worldwide.

© Springer International Publishing Switzerland 2017

M. Carroll, *Earths of Distant Suns*,

DOI 10.1007/978-3-319-43964-8_9

219

Others complain that we can never hope to understand what an alien mind would try to say to us. But as we have seen, if a message is received in the radio spectrum, our civilization shares with them the commonality of radio science, mathematics and physics.

Still others warn that throughout our history, the clash of cultures has resulted in destruction of the younger, less advanced one. Why, they say, should we advertise our existence? As astrophysicist Neil DeGrasse Tyson said in an interview with the *Business Insider*, "We don't give our address to members of our own species whom we don't know. So, the urge to give our home address to aliens? That's audacious." But as SETI experts have pointed out so clearly, we've been advertising ourselves since the broadcast days of *The Ed Sullivan Show*[2] and *Gilligan's Island*. If they are there and listening, they know we are here already. And far from the evil extraterrestrials of tales like *War of the Worlds* or *Independence Day*, alien biology will be so different from ours that they will not want us as an appetizer in their lunchbox. Rather, they may not be interested enough to pick up the phone (see Chap. 6, Fermi Answer #3).

If we can engage in some kind of dialog, will we be able to understand each other even on a fundamental level? In her 1996 masterpiece *The Sparrow*,[3] Mary Doria Russell pens the provocative story of a Jesuit priest sent to a planet in the Alpha Centauri system after Earth's Arecibo station picks up music coming from there. The priest interacts with two cultures there, and in a series of misunderstandings, the mission ends in tragedy and murder. The cultural and linguistic missteps are consistent with historical culture clashes, and tell a cautionary tale about our interaction with an alien race.

Polish author Stanislaw Lem takes the theme even further, suggesting that alien thought will be so very different from ours that any communication will prove impossible. In his remarkable novel *Solaris* (1962), his central message is clear: humans will never understand the alien mind. His book envisions an alien entity, ensconced in a global ocean, which probes human minds in orbit as they probe it. The two life forms are so different that they can never have a meaningful dialog. But others, including Carl Sagan, feel that we will be able to find common ground, perhaps based on the mathematical principles held so dear by SETI researchers.

---

[2] Which means that any sentient beings out there who have monitored our *Ed Sullivan Show* broadcasts have heard the music of the Beatles and the Beach Boys. We have put our best foot forward.

[3] Ballantine Books, 1997.

Life's Rich Pageantry

As our search continues, we wonder not only what form their communication will take, but what wondrous forms life itself might take beyond the shores of terra firma. Nineteenth-century astronomer Camille Flammarion said, "The eminent men of all ages who have been versed in the operations of nature have also been profoundly impressed by its prodigious fecundity and have understood the insanity of those who would limit that fecundity to our abode alone." He went on to warn that life may take on forms far different from Earthly ones. He admonished those who assumed that intelligent human-like beings lived among the stars, accusing them of "hurling a gross insult in the shining face of the infinite Power who fashions the worlds."[4]

Flammarion's caution was ignored by writers of such modern movies as *Men in Black*, which populates Earth with a host of aliens, many of which take on very human characteristics. At one point, one character reveals to another that Dennis Rodman is an alien. The other replies, "Not a very good disguise."

Loran Eisley said, "So deep is the conviction that there must be life out there beyond the dark, one thinks that if they are more advanced than ourselves they may come across space at any moment, perhaps in our generation…in the nature of life and in the principles of evolution we have had our answer. Of men elsewhere, and beyond, there will be none forever."[5] Later, Eisley adds, "Life may exist in yonder dark, but it will not wear the shape of man."

According to Douglas Adam's *Hitchhiker's Guide to the Galaxy*, the third most intelligent life form on Earth is the primate, followed by the more intelligent dolphin[6] and— in first place—the laboratory mice who are actually carrying out all the intelligence tests on humans. His subtle message has more to do with human hubris than anatomy and intelligence, but a secondary point is that we do not know what forms intelligent life may take.

Life may, indeed, take on many different visages throughout the universe. Not all life engages with its environment with the same senses of touch, sight or smell. If a life form relates to reality on such different terms, how would we make contact? Mr. Spock used the Vulcan mind meld on the Horta. When it comes to cross-species communications, human techniques are more limited.

---

[4] Crowe, M. J., *The Extraterrestrial Life Debate, Antiquity to 1915* (Notre Dame, 2008).

[5] Eisley, Loren, *The Immense Journey* (Random House, 1957).

[6] According to the *Hitchhiker's Guide*, humans misinterpreted the dolphins' warnings of Earth's impending doom as "amusing attempts to punch a football or whistle for tidbits." This is a good cautionary tale for SETI researchers.

We must find a celestial lingua franca, and that universal language may be mathematics. If, as the SETI experts claim, mathematics and science will be our first common ground, there will undoubtedly be other obstacles. What do we base our assumptions upon in our mathematics? We use a base ten system because we have ten fingers—"digits"—on our hands. But the Babylonians used a base six system, and computer programmers are familiar with the hexadecimal (base 16) system as well.

Shostak does not see this numerical issue as a stumbling block. "If our technology is dependent on mathematics, and our math is based on a numbering system from one to ten because that's the number of digits we have and they may have something different, would that change anything? I suspect not. If our numbering system went from one to twelve, we would have still figured out Maxwell's equations."

Visualizing World SETI

Many see the discovery of extraterrestrial civilizations as a unifying factor for Planet Earth. While addressing the United Nations General Assembly in 1988,[7] then-president Ronald Reagan made the case this way: "I occasionally think how quickly our differences worldwide would vanish if we were facing an alien threat from outside this world. And yet, I ask you, is not an alien force already among us? What could be more alien to the universal aspirations of our peoples than war and the threat of war?"

Reagan was speaking at the height of the Cold War, at a time when the world stood on the brink of nuclear holocaust, with literally thousands of nuclear warheads pointed at each side, one hemisphere to the other. Nuclear was not the only threat to humanity. Advances in chemistries and manufacturing meant pollutants were reaching the highest levels in history. Conspicuous consumption drained natural resources. Species were becoming extinct at alarming rates due to this pollution and the destruction of wilderness areas.

Against this backdrop, our planet faced what many saw as insurmountable challenges that might well lead to the mass extinction of much life on Earth. And while many of the walls have come down and disarmament is under way globally, pollution still dogs us, waste runs rampant, and a new generation of nations are arming themselves with nuclear weapons. Old Cold War stresses still haunt us with the specter of self-destruction. But at the same time, we have, at this flickering moment in history, learned how to leave our

---

[7] Presidential address to the United Nations General Assembly, September 21, 1987.

cosmic nest. We have, as a race, spread our wings and taken the first fledgling steps into interplanetary space.[8] Our engineers have discovered ways to harness renewable energy, culling power not from the burning of fossil fuels but from the wind and sunlight and seas of our planetary surroundings. We are exploring both edges of our technological sword. We stand not on a precipice but at a crossroads. Is it possible for an intelligent species to survive such a crossroads? The discovery of advanced civilizations in the cosmos will give us an affirmative, positive, hopeful answer.

Many thinkers believe advanced alien life, if it has survived the trauma that Earth's civilizations now face, will have learned to become beneficent. Carl Sagan believed the great expanses between distant suns were providential, providing a sort of natural quarantine. In his masterpiece *Pale Blue Dot*, he said, "The quarantine is lifted only for those with sufficient self-knowledge and judgment to have safely traveled from star to star."

Over the course of the history of the universe, however, is it possible that Earthlike planets formed in the earlier epochs, or was the universe too different to form ancient Earths? One of the most astounding planetary systems we have found is Kepler 444. The M dwarf's confirmed planets consist of five small, sub-Earth-sized rocky worlds huddling in tight orbits very near each other. The Kepler 444 system is the most compact yet found. All of its planets circle their star in orbits lasting less than 10 days, far too close to their sun to be within the habitable zone. They are the oldest known terrestrial planets.

NASA's Elisa Qunitana marvels at how ancient the system is. "Kepler 444 made me think not only of its special scale, but the time scale as well. How many billions of years have there been Earths born and died and born and died?" Quintana says the system shows that from the beginning of the galaxy there were planets. "These worlds are 11 billion years old." When the planets formed, our universe was only 20 % as old as it is today. The length of time available for civilizations to arise makes SETI even more reasonable.

Astronomer Jill Tarter observes, "SETI is sexy, it's of interest to many people, particularly young people. If we can engage the globe—humans everywhere—in thinking about this, it has the potential for changing their perspective. If you get serious and think about SETI for a while, you are inevitably forced to take a much broader view. When thinking in this way, you are forced to the conclusion that when compared to something else out there, we are all the same. I think that is the most important perspective-changing step that humans can take: to understand that we are all Earthlings. We need to be able to operate globally as a way of husbanding this planet and our resources.

---

[8] And, if you count the Voyagers and New Horizons, into interstellar space.

If SETI can get global cooperation started on a very small scale with this kind of searching, maybe that's a training ground for taking on other big problems that don't respect national boundaries."

Why Bother?

If the chances are great that we will never make direct contact with other sentient beings, should we bother to search? Why?

Many feel the answer is yes. First, if technological beings can avoid sentient suicide and survive extinction events, they will have had the chance to advance far beyond our technology (especially if they have been around as long as the Kepler 444 system). And while communication between stars is apparently a one-way conversation, other cultures might well send something like an encyclopedia or a cultural portfolio to us. We could probably learn a thing or two of value from them.

Second, we would have the assurance that civilization can, in fact, survive to a more mature level than we have achieved. This is significant both philosophically and practically.

Third, perhaps we can finally answer the age-old question, *Are we alone in the universe?* Questions like that tend to give one perspective, to put us in our place. In Arthur C. Clarke's story *Childhood's End,* the people of Earth gave up learning anything because of the great chasm between what the aliens knew and what we might discover on our own. Their attitude became, *why bother?* But the discovery of cultures on distant Earths could also invigorate our species. New information stimulates new ways of thinking. In the Renaissance, artists, philosophers and inventors rose from medieval cultures. The advent of open borders and exploration combined with the mixing of cultural ideas, engendering the European "rebirth," or Renaissance. Many SETI experts assert that the revelation of another civilization would undoubtedly bring unique perspectives and new ways of thinking to our own cultures.

Futurist author Nikos Prantzos says, "Any encounter with an extraterrestrial civilization would dramatically alter our approach…Personally, I think it will be a lesson in humility. If on the other hand we find that we are the only form of intelligence in the galaxy, it will be our duty to preserve nature's uniquely successful experiment and spread it across the whole universe."

The knowledge that other civilizations have not self-destructed, have thrived, and have advanced beyond what we have, provides us with an optimistic view of our possible fate. Many feel that the lack of discovery so far should not dissuade us, but that we should press on, bringing to bear our resources and creativity to the great search. The discoveries of distant Earths may well transform our outlook and behavior as one of many races of sentient beings among the stars.

SETI All Grown Up

Technological advances have brought maturity to the search for extraterrestrial intelligence. These advances have been made in the hardware used and the software that drives it. SETI engineers estimate that our telescopes and receivers are within a factor of two of advancing as far as physically possible. Now, it is up to the computers, says Jill Tarter. "Whether you're looking for radio or optical signals, you have to search through vast amounts of parameter space, and you need enormous computing resources to do it. Now we are adding exponentially increasing amounts of computing so that we can use the tools systematically to observe large chunks of space. If you just pop your head up and look around, you really can't make significant statements of absence of life, or probabilities. If you use these tools systematically, we can whittle down the space where signals could exist and make some statements."

Within the scientific community, attitudes toward SETI are changing, says NASA Ames' Tom Barclay. "In the eighties, say, the idea of searching for life was different. Scientists who spoke of it weren't taken too seriously. Even among the older scientists, there is a desire to be extremely careful with language when they're talking about things like this. But now, among the younger scientists, it's fine to say 'I want to find life. I want to find aliens.' You're not ridiculed. It's not a ridiculous thought. We're finding places that look like Earth out there. We've got hard data that places like Earth are probably common."

In fact, two decades ago we knew only of the planets in our own Solar System. Twenty years from now, many people will have grown up with exoplanets as a common aspect of the known cosmos. Astronomers have progressed from knowing very few planets to knowing that planets are everywhere, and the timescale of this change has been a very few years. Researchers are now at the point where they are predicting, with confidence, what the next discoveries will be and when they will happen. Much of their confidence comes from the future telescopes on the drawing board and nearing the launch pad, and from the recent track record of discoveries. Barclay comments, "Many of us study exoplanets because we want to know: are we alone? That makes the study of this the most exciting field imaginable, because that's the most exciting question there is."

Culture Clash

As the discovery of exoplanets makes its way into our culture, it affects not only our movies but our music, our art and our literature. Like other artists, poets have been influenced by the vast horizons offered within the search for extraterrestrial companions. Polish Nobel laureate Wislawa Szymborska penned a work entitled *The Ball*:

> As long as nothing can be known for sure (no signals have been picked up yet),
>   as long as Earth is still unlike the nearer and more distant planets,
>   as long as there's neither hide nor hair of other grasses graced by other winds,
>   of other treetops bearing other crowns, other animals as well-grounded as our own,
>   as long as only the local echo has been known to speak in syllables,
>   as long as we still haven't heard word
>   of better or worse mozarts,
>   platos, edisons somewhere,
>   as long as our inhuman crimes
>   are still committed only between humans,
>   as long as our kindness
>   is still incomparable,
>   peerless even in its imperfection,
>   as long as our heads packed with illusions
>   still pass for the only heads so packed,
>   as long as the roofs of our mouths alone
>   still raise voices to high heavens –
>   let's act like very special guests of honor
>   at the district-firemen's ball
>   dance to the beat of the local oompah band,
>   and pretend that it's the ball
>   to end all balls.
>   I can't speak for others – for me this is
>   misery and happiness enough:
>   just this sleepy backwater
>   where even the stars have time to burn
>   while winking at us

Other writers wonder about the interface between religious faith and the discovery of alien races, as Wells referred to in his opening of *War of the Worlds*. Speaking of the "more ancient worlds" (as Mars was then considered), he said: "At most terrestrial men fancied there might be other men upon Mars, perhaps inferior to themselves and ready to welcome a missionary enterprise."

To the Christian, Mormon or other believer who embraces the idea of evangelism, our role with other races becomes problematic. Do we become the mouthpiece of a God who sacrificed himself "once for all" in order to bring salvation to an alien civilization? Ray Bradbury explored this concept in his short story "The Fire Balloons." Priests who come to Mars are confronted by peaceful aliens who no longer have physical bodies. What, to them, is sin? Do they face temptation of a completely different kind than that common to humankind?

Like many scientists and philosophers, poet Alice Meynell saw no conflict between faith and the astronomical sciences. Exploring the premise in her poem "Christ in the Universe," she wrote:

> With this ambiguous earth
>   His dealings have been told us. These abide:
>   The signal to a maid, the human birth,
>   The lesson, and the young Man crucified.
>
> But not a star of all
>   The innumerable host of stars has heard
>   How He administered this terrestrial ball.
>   Our race have kept their Lord's entrusted Word.
>
> Of His earth-visiting feet
>   None knows the secret, cherished, perilous,
>   The terrible, shamefast, frightened, whispered, sweet,
>   Heart-shattering secret of His way with us.
>
> No planet knows that this
>   Our wayside planet, carrying land and wave,
>   Love and life multiplied, and pain and bliss,
>   Bears, as chief treasure, one forsaken grave.
>
> Nor, in our little day,
>   May His devices with the heavens be guessed,
>   His pilgrimage to thread the Milky Way
>   Or His bestowals there be manifest.
>
> But in the eternities,
>   Doubtless we shall compare together, hear
>   A million alien Gospels, in what guise
>   He trod the Pleiades, the Lyre, the Bear.

O, be prepared, my soul!
  To read the inconceivable, to scan
  The myriad forms of God those stars unroll
  When, in our turn, we show to them a Man.

Dreams of Distant Earths

If the galaxy is teeming with life, we wonder what alien cultures might dream about and aspire to. As we wonder, our minds turn to the ancient cultures of our own world, cultures quite alien to the modern mind. We find pictographs and petroglyphs in caves and across the cliff faces, spanning from 25,000–year-old ice-age sites in Europe to sites in the American desert southwest dating back as far as 10,000 years. These enigmatic images, many not yet understood by anthropologists, conjure the kind of challenge that we will need to face if we encounter beings from other star systems. What were our paleo artists trying to do? We wonder if the dancing figures, elegant bison and inscribed spirals on cave walls equate to anything that we would call art (Fig. 9.1). Were they after a simple depiction of their surroundings, or did their art reflect some magical or spiritual component?

Astronomer/artist William K. Hartmann asks, "How did the people who made them *think about them*? Shamanistic activity? Harmless fun? Marking their land? Did they take their 10-year-old boys back to the image chambers to talk about hunting and big animals and warfare and fear? Did women make some of the paintings? All of them? Or did just shaman-artists go back there? And are all artists shamans of a sort?"

The images of our ancestral cultures, which arose on the same planet we did, baffle us. How much more will we be puzzled by what we find out in the cosmos? If we are at the threshold of technological advancement that will enable us to engage with that life, we must also be at the dawn of new revelations about not only alien cultures but our own (Fig. 9.2).

Engaging with the Others

Among the most ancient of human writings, we find the story of Babel, in which the Creator of the universe disciplines his presumptuous children by confusing their speech, fracturing the world's population into groups of many disparate languages. Perhaps, in our efforts at communicating with civilizations

Fig. 9.1 Experts still have only a partial understanding of the mysterious petroglyphs in the deserts of the southwestern United States. (Painting © Erica McGinnis, used with permission)

on Earths of distant suns, we will find common language and universal ways to communicate, not only with alien beings but with fellow humans on our own planet. And as Babel was a monument to hubris, we can hope that SETI

Fig. 9.2 The Kepler 444 system has been around for billions of years, plenty long enough for advanced civilizations to have come and gone. Here, planet 444f simmers under the heat of its sun just a tenth of an AU away. Its atmosphere long dissipated, all that is left behind are remnants of a civilization's outpost. Planets Kepler 444e (*left*) and 444d transit across the star's face. (Art © Michael Carroll)

will ultimately become a vehicle from which humanity learns humility, a teacher to put us in our place in a cosmos thriving with life and civilization. That cosmos may await us as we stand, today, at the beginning of our search for life among Earths of distant suns.

Index

© Springer International Publishing Switzerland 2017
M. Carroll, *Earths of Distant Suns*,
DOI 10.1007/978-3-319-43964-8